James Caleb McIntosh

The Prophetic Voice of God

The Unparalleled Fires, Floods, Hurricanes, etc. of the Years 1866, 1867 and 1868

James Caleb McIntosh

The Prophetic Voice of God
The Unparalleled Fires, Floods, Hurricanes, etc. of the Years 1866, 1867 and 1868

ISBN/EAN: 9783743422094

Manufactured in Europe, USA, Canada, Australia, Japa

Cover: Foto ©berggeist007 / pixelio.de

Manufactured and distributed by brebook publishing software
(www.brebook.com)

James Caleb McIntosh

The Prophetic Voice of God

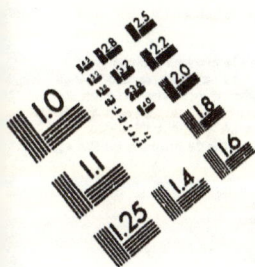

**IMAGE EVALUATION
TEST TARGET (MT-3)**

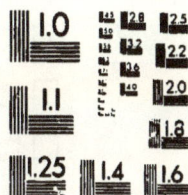

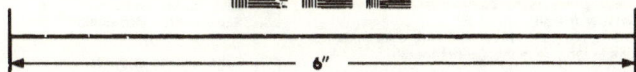

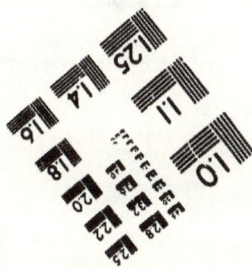

|← 6" →|

Photographic
Sciences
Corporation

23 WEST MAIN STREET
WEBSTER, N.Y. 14580
(716) 872-4503

The copy filmed here has been reproduced thanks to the generosity of:

D. B. Weldon Library
University of Western Ontario

The images appearing here are the best quality possible considering the condition and legibility of the original copy and in keeping with the filming contract specifications.

Original copies in printed paper covers are filmed beginning with the front cover and ending on the last page with a printed or illustrated impression, or the back cover when appropriate. All other original copies are filmed beginning on the first page with a printed or illustrated impression, and ending on the last page with a printed or illustrated impression.

The last recorded frame on each microfiche shall contain the symbol → (meaning "CONTINUED"), or the symbol ∇ (meaning "END"), whichever applies.

Maps, plates, charts, etc., may be filmed at different reduction ratios. Those too large to be entirely included in one exposure are filmed beginning in the upper left hand corner, left to right and top to bottom, as many frames as required. The following diagrams illustrate the method:

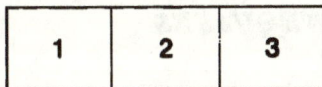

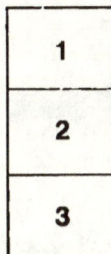

L'exemplaire filmé fut reproduit grâce à la générosité de:

D. B. Weldon Library
University of Western Ontario

Les images suivantes ont été reproduites avec le plus grand soin, compte tenu de la condition et de la netteté de l'exemplaire filmé, et en conformité avec les conditions du contrat de filmage.

Les exemplaires originaux dont la couverture en papier est imprimée sont filmés en commençant par le premier plat et en terminant soit par la dernière page qui comporte une empreinte d'impression ou d'illustration, soit par le second plat, selon le cas. Tous les autres exemplaires originaux sont filmés en commençant par la première page qui comporte une empreinte d'impression ou d'illustration et en terminant par la dernière page qui comporte une telle empreinte

Un des symboles suivants apparaîtra sur la dernière image de chaque microfiche, selon le cas: le symbole → signifie "A SUIVRE", le symbole ∇ signifie "FIN".

Les cartes, planches, tableaux, etc., peuvent être filmés à des taux de réduction différents. Lorsque le document est trop grand pour être reproduit en un seul cliché, il est filmé à partir de l'angle supérieur gauche, de gauche à droite, et de haut en bas, en prenant le nombre d'images nécessaire. Les diagrammes suivants illustrent la méthode.

1	2	3

1
2
3

THE

Prophetic Voice of God,

OR

SOME ACCOUNT

OF THE

UNPARALLELED FIRES, FLOODS, HURRI-

CANES AND FAMINE, CRIMES, PLAGUES

AND EARTHQUAKES

OF THE YEARS 1866, 1867, AND 1868.

SELECTED AND COMPILED BY

JAMES CALEB M'INTOSH, BAYFIELD,

COUNTY OF HURON, ONTARIO,

JANUARY, 1869, A. D.

THE

Prophetic Voice of God,

OR

SOME ACCOUNT

OF THE

UNPARALLELED FIRES, FLOODS, HURRI-

CANES AND FAMINE, CRIMES, PLAGUES

AND EARTHQUAKES

OF THE YEARS 1866, 1867, AND 1868.

SELECTED AND COMPILED BY

JAMES CALEB M'INTOSH, BAYFIELD,

COUNTY OF HURON, ONTARIO,

JANUARY, 1869, A. D.

PRINTED AT THE SIGNAL OFFICE, GODERICH.

To the Reader.

Tho following book will be found but a mere summary. A full account of each of the subjects contained in it would swell to volumes. Neither can it be expected in every point to be correct, but this I do say, that I have followed my information without any deviation on my part. Hence, I shall be content to have it perused by a discerning public.

———————

CONTENTS:

CHAPTER I.

STATISTICS.

From the New York Tribune of 1866.

" THE POPULATION OF THE GLOBE.—There are on the globe about 1,288,000,000 souls, of which 369,000,000 are of the Caucasian race; 552,000,000 are of the Mongol race; 190,000,000 are of the Ethiopian race; 176,000,000 are of the Malay race; 1,000,000 are of the Indo-American race. There are 3,648 languages, and about the complete number of sects 666. The real mark or number of the Beast " The yearly mortality of the globe is 33,332,333 persons. This is at the rate of 91,554 per day; 3,730 per hour; 60 per second. To each pulsation of our heart marks the death of some human being. The average of life is 33 years. One fourth of the population die at or before the age of seven years, one half at or before 17 years. Among 10,000 persons one arrives at the age of 100 years, one in 500 attains the age of 90, and one in 100 lives to the age of 60. Married men live longer than single ones. In 1,000 persons 65 marry, and more marriages occur in June and December than in any other two months in the year. One-eighth of the whole population is military. Professions exercise a great influence on longevity. In 1,000 individuals who arrive at the age of 70 years, 42 are priests, orators, or public speakers, 40 are agriculturists, 33 are workmen, 38 soldiers or military employees, 29 advocates or engineers, 27 professors, and 24 doctors. Those who devote their lives to the prolongation of that of others die the soonest. There are who profess christianity 335,000,000; there are 5,000,000 Israelites, 60,000,000 of the Asiatic religion; there 160,000,000 Mahommedans; there are 200,000,000 Pagans. Them that profess the Catholic belief number 170,000,000. The Greek church numbers 75,000,000, and 80,000,000 profess the Protestant faith. In forty years in the United States, Great Britain and France, from 1820 to 1860, the population is shown to have increased as follows :

	1820.	1860.
United States, ...	9,638,191	31,445,080.
France, ...	30,461,875	36,755,371.
Great Britain, ...	20,892,670	28,887,587.
1 England and Wales,....	11,999,322	20,001,725.
2 Scotland, ...	2,091,521	3,061,820.
3 Ireland, ...	6,801,827	5,764,543.

Dividing three millions into city and country population, the same forty years, outside of the principal cities of Great Britain, viz: London, Manchester, Liverpool, Leeds, Bristol, Birmingham, Glasgow and Dublin, and in France, outside of Paris, and in the United States, outside of its fifty principal cities, the result is :

	1820.	1860.
In Great Britain, ...	18,641,733	23,736,405.
In France, ...	29,701,875	35,088,030.
In United States, ...	9,068,181	27,354,287.

Whereas the growth of population during the same forty years in the eight above named principal cities of Great Britain, in the capital city of France, and in the fifty chief cities of the United States, is as follows :—

	1820.	1860.
Great Britain's 8 cities,	2,250,937	5,151,192.
France's Capital city,.....	760,000	1,667,841.
United States' fifty cities,	570,010	4,090,793.

In Great Britain the increase ratio of the country population is about 4 to 5½, while, of the city population it is 4 to 9. In France the increase ratio of the country population is 4 to less than 5, while of the population of Paris it is from 4 to 9. In the United States the increase ratio of the country population for the same period, although rising from 4 to 12, is outstripped by the increase of the population in cities, which is from 4 to nearly 29. But not only do cities outstrip the country in their growth, but great cities outstrip smaller cities. In Great Britain—Manchester, Liverpool, Leeds, Bristol and Birmingham, have increased in their aggregate population from 539,060 in 1820, to 1,651,075 in 1860. London, in 1820, had 1,373,947 inhabitants; the same ratio of increase as the five cities above, would give London in 1860 about 2¼ millions of human beings, but it has gone beyond that mark by half a million, just enough to make a city of the size of Manchester the next largest in the Kingdom, and London in 1860 had a population of 2,750,000. Glasgow, the chief city of Scotland, has increased three fold in the same period, far surpassing its rivals, while the country has increased fifty per cent. Dublin has risen from 185,000 to 250,000 steadily, despite the fluctuations of population. In the United States there is uniformly more rapid concentration of population in the great central cities, than in the cities at large, and is more strikingly manifest, as thus:

	1820.	'1860
Forty-eight principal cities,	439,129	3,009,878.
Seven larger cities, ...	266,304	1,452,521.
New York, Brooklyn, Williamsburg, Jersey City,	130,671	1,110,410.

RAILROADS IN FRANCE.—The total length of railroads in operation January 1st 1865, was 8,113 miles, and concessions had been granted for 3,304 miles, making a total of 11,417 miles completed, in progress and projected. The amount of money actually expended on these enterprises to date, was $1,300,000,000, and there remained to be expended on the roads in progress and projected an additional sum of $570,000,000, which makes a total of $1,870,000,-000, or about $150,000 per mile. The tunnels on all the railroads in the Empire are 366 in number, and would, if combined, measure 377 leagues in length. The largest is that of the North near Marsailles, on the Lyons Railroads, which cost $2,100,000 ; and that of Blaisy, on the same line cost $1,600,000 ; and that of Credo, between Lyons Geneva, $1,300,000. The entire cost of the tunnels, bridges and viaducts on the various French Railroads amounts to $86,536,390.

"About the year 14 of the Christian era, the annual product of gold was $5,000,000 ; in 1492 it was only $250,-000 ; in 1863 it was $285,000,000 ; and in 1864 $240,000,000. In the year 14 also the gold and silver in existance is estimated at $1,327,000,000, and in 1862 at $10,562,000,000. The whole amount of gold and silver obtained from the earth from the earliest periods to the present time is estimated at $21,272,000,000.

FACTS ABOUT THE BIBLE.

The Scriptures have been translated into 148 languages and dialects, of which 121 had prior to the formation of the British Foreign Bible Society ever appeared. And 25 of those languages existed without an alphabet, in an oral form. Upwards of 43,000,000 of these copies of God's word are circulated among not less than 600,000,000 of people. The first division of the Divine word into chapters and verses is attributed to Stephen Langton, Archbishop of Canterbury, in the reign of King John, in the latter part of the twelfth century or beginning of the thirteenth. Cardinal Hugo, in the middle of the thirteenth century, divided the Old Testament into chapters as they stand in our present translation. In 1661, Athias, a Jew of Amsterdam, divided the sections of Hugo into verses—a French printer had previously (in 1561) divided the New Testament into verses as they now are. The Old Testament contains 39 books, 929 chapters, 23,214 verses, 592,-439 words, 2,738,100 letters. The New Testament contains 27 books, 260 chapters, 7,950 verses, 182,253 words, 933,-380 letters. The entire Bible contains 66 books, 1,139 chapters, 31,175 verses, 774,692 words, 3,565,489 letters. The name of Jehovah, or Lord, occurs 6,855 times in the Old Testament. The word "and" occurs in the Old Testament 35,543 times. The middle book of the Old Testament is Proverbs. The middle chapter is the 29 of Job. The middle verse is the 2 of Chronicles, 29th chapter, 17th verse. The middle book of the New Testament is 2nd Thessalonians. The middle chapters are Romans 13 and 14. The middle verse is Acts ii, 7. The middle verse in the Bible is Psalms cxviii, 8. The middle line in the Bible is 2nd Chronicles i, 16. The 19th chapter of 2nd Kings, and Isaiah 36 are the same. In the 27th verse of the 7th Ezra, are all the letters of the alphabet, I and J considered as one. The Apocrypha so called, is as canonical as the rest of the Bible, and is bound between the Old Testament and New, and contains 14 books, 183 chapters, 15,031 verses, 153,185 words. The preceding facts were ascertained by a gentleman in 1718, also by an Englishman residing at Amsterdam, 1772, and is said to have taken them each nearly three years in the investigation. The thought that an eternity of bliss depends upon the purity of a few years of earthly existence, is an overwhelming thought.

Then, how great is the inducement to study truth, and cultivate virtue.

How happy every child of grace,
Who knows, his sins forgiven ;
This earth he says, is not my home,
I seek my place in heaven.
A country far from mortal sight,
Yet oh, by faith I see ;
The land of rest, the Saint's delight,
The heaven prepar'd for me.

CHAPTER II.

PLAGUES.

The definition of the word plague, is pestilence, or that which cuts off any thing that God created. First, then, the cattle plague. From the Lendon *Times*, December 65 :—" Official returns for the week ending December 30, show a large increase, in fresh cases of the cattle disease. For the previous week the number was 6,256, and in the last week of December 65, it was 6,693 The number in Yorkshire was 1,456 ; in Scotland 1,975. The United States Consul, writing to the State Department from Manchester, under date of December 9, 65, says : Since the date of my last despatch on the subject of the cattle plague, the deaths have rapidly increased. The whole number of cases reported to the authorities up to the close of last week is 39 000 ; of these 19,950 died, and 10,700 were killed while suffering with the disease. The number of cases reported last week was nearly 4,000. The exact figures are, for the week ending November 18, 2,069 ; for the week ending November 25, 3,610 ; for the week ending December 23, 3,828 ; for the week ending January 27, 66, show the number attacked to be nearly 12,000, against 10,000 the previous week. From the same authority we regret to notice the alarming increase in the cattle plague. During last month, the attacks have risen from 14,000 to upwards of 27,000, and the deaths are daily increasing."

" Statistical office of the Veterinary Depot, Privy Council office, December 20, 1866:—The number of animals attacked since the commencement is 253,791, and 52,496 healthy cattle have been slaughtered to prevent the spread of the disease." •

London *Times*, Nov., 1866:—" There can be no longer a doubt about the formidable growth of the cattle plague, while the deaths from plague and pole-axe together, were only about 12,000. This gave a mortality of 700 a week upon an aggregate stock of horned cattle estimated at 7,-000,000, and it was asked accordingly whether such a loss would justify the application of such a remedy as was proposed. Cost of the cattle plague to England : It has been estimated that the actual loss in money by the cattle plague has been $17,865,000. Of the cattle attacked by the plague 210,000 are dead, and at $60.00 each—for the old estimate at $50.00 a head does not adequately represent the prices realized during the last few years—their value is $12,600,000. The Government have slaughtered 53,000 head, to prevent the spread of the disease; to these add a much greater number, say 100,000 animals, butchered in an unripe state, making about half price at market; that is 153,000 altogether, at $15 00 each, losing $4,590,000. Then 24,000 of the attacked recovered, while 11,000 are unaccounted for ; say that the decrease in value upon these was $15.00 a head on 45,000 animals, losing $675,-000 more." I shall now leave the reader a few moments to reflect upon the sad account we had from the London *Times*, of the cattle plague in England, and give a brief account of the cattle plague in Holland.

Brussels, December 14, 66.—The Belgian *Moniteur* publishes the following particulars concerning the cattle plague in Holland :—" The cattle plague appears to be making dreadful ravages among the cattle in Holland. The number of fatal cases do not cease to increase and if the progress observed to have been made by the disease since November last continue, the losses of the Dutch farmers will soon exceed those of the English cattle owners, at the time when the plague was most violent. According to the official reports the number of cases among cattle were for the weeks ending November 3, 1,443 ; 10th, 1,551 ; 17th, 1,595 ; 27th, 3,257 ; and December 4th, 7,162. The last number is more than double that which was recorded when the epidemic was at its worst in December, and everything tends to show that it does not indicate the greatest height of the disease. The cattle plague was especially virulent in the provinces of Utrecht and Southern and Northern Holland, but it has also shown itself in Friesland and Overyssel, and has latterly attacked many parishes of Guelderland and North Brabant. Belgium having such an extended and frequent relations with Netherlands is especially menaced by these circumstances. She will only succeed in keeping herself free from contagion by never for a moment relaxing the most rigorous vigilance and circumspection. Let the reader only remember, and add to the sad tale already told, the hundreds of thousands died from the same cause in the United States, and other places not adjacent to us ; the spirit of inquiry should rap at the door of reason, and ask : What is all this for ? When God was about delivering the children of Israel from Egyptian bondage, he sent his plagues in a similiar manner upon their oppressions. See Exodus ix, 1, 6. Comment is useless."

THE GRASSHOPPER SCOURGE.—From the Marysville *Enterprise*, 1867 :—"The Kansas farmers in Brown County and adjacent territory, appear to have been lately subjected to a plague similiar to those inflicted on Pharaoh. See Exodus x, 5, 6."

The *Enterprise* says :—"The obstinate grasshoppers appeared in countless numbers, covering a tract twelve miles in width, consuming almost all vegetation. They alighted upon fields, gardens, fruit trees, and everything green or eatable; and like a march of two hundred and fifty army corps, devoured everything they touched, This whole country has been taken by them, and the rear guard is still with us, guarding what vegetables and green leaves the army has left. Farmers are seriously alarmed lest the corn will be totally destroyed."

Reader, remember also, the suffering at Red River in 1868, through the plague of grasshoppers, and how they swept everything green in their path, in countless numbers; and how different is our situation, in the County of Huron in December 1868, when I am writing this book. Flour with us, is only worth four dollars and fifty cents per barrel, when at the same time at Fort Garry, flour is $28.00. Reader, take my opinion for it, our time for suffering and affliction is at hand, and we shall not escape.

From the Weekly *Globe*, of 66 :—"The barque "Paiz," has arrived from Hong Kong. Fifty-two per cent, of the British garrison at Hong Kong had died in nine months, from some new and strange disease."

From the New York *Tribune*, of 66 :—"The total number of deaths from cholera in Paris, from its first appearance to the 30th of November, was six thousand and seventy-seven."

From the London *Times* —Progress of cholera in Europe in 1865-6 :—"The Registrar-General of England, in a supplement to his last weekly report, has published a series of authentic official returns, recording some valuable information relating to this plague of the 19th century, in the cities and towns of Europe. The French returns show that in Paris the epidemic of 1865 reached its maximum in October, in which month 4,663 deaths were recorded. In the first six months of 1866, 69 deaths occurred, but in July the last month for which returns have been published, the deaths suddenly rose to 1,743. The proportional number of deaths by cholera to every 10.000 of the population of Paris in '65, was 39 ; in the first seven months of 66, it was 11. In London the deaths by Cholera in '66 were in the proportion of 18, in Liverpool, of 36 to 10,000 living. In Italy the epidemic began in the 25th of June, 1865, in the province of Tourin, and destroyed 12,901 lives during that year, or to every 10,000 of the population living in the 35 provinces, and the 349 communes that were attacked by cholera 35 deaths occurred. It appears that in Italy the town population has suffered less severely than that of the country, the number of deaths in 10,-000 living, being 38 in the former and 50 in the latter case. In Naples, 2,301 deaths are recorded in '65, out of 446,931 inhabitants, being in the proportion of 52 deaths by cholera in every 10,000 living. In Vienna the returns, date from the 11th of August to the 10th of November 1866, the ratio was 51. In seven Belgian towns, comprising Antwerp, Brussels, Bruges, Ghent, Mons, Liege and Namur, no less than 11,771 deaths occurred from May 1, to October 15 of 1866, out of a population of 553,377, or the deaths by cholera were in the proportion of 186 to 10,000 living. In Brussels the proportion was 164. In Holland 18,547 deaths occurred in 1866. Taking 15 Dutch cities and towns including Amsterdam, it appears that 8,872 deaths

by cholera were recorded in the five months from June
to October '66, being in the proportion of 107 deaths to
every 10,000 living. In Amsterdam the ratio was 42, while
in Utrecht it was 271.   Norway, it appears, has suffered
but slightly from the epidemic in 1866, only 48 deaths out
of 1,000,000. Progress of the Cholera. Fearful cholera
panic at Madrid—five hundred deaths per day. The London *Gazette* says, the number of deaths are over five hundred per day, and the population is under half a million ;
this is a mortality more than 15 times that of London at
the present time, 1866. The consequence is that the
population are leaving the capital in thousands. Indeed,
so violent is the stampede that the people rush into the
trains without taking their tickets, in fear that they may
be left behind, and it has been found necessary to place
troops at the railway stations in order to preserve order.
The cholera has made greater ravages in the convents, and
this is not surprising, for the nuns who die are buried beneath the buildings, after having been laid out in the
chapels for 36 hours. The Queen has been anxiously
desirous of returning to Madrid, in order to restore the confidence of her panic-stricken subjects. But her Ministers,
having regard to the fact that she is enciente, and the political complications that would arise should she fall
a victim, have dissuaded her, and have thereby given oc-
casion to the opposition journals to compare unfavourably
to the former the conduct of Queen Isabella with that of
Pedro V, of Portugal, who during the plague in Lisbon,
visited the hospitals personally. The cholera has also
spread to Portugal, having appeared in the town of Elvas,
which is just beyond the Spanish frontier, and is about ten
miles from Badajos. If, along with the foregoing, we
take into account the cholera in this Province, Canada,
and also in the United States, with many other places, that
has been visited by the same scourage. I think that portion of our Redeemer's language, is having its fulfilment.
See Matthew xxiv, 7. " And there shall be famines, and
pestilences, and earthquakes, in divers, or, different places."
And those he gave, as tokens of his coming.

From the New York *Tribune* :—"The Drought in Australia.—At the departure of the last mail from Australia
rains had fallen, and it was thought the drought was
breaking up, but it still continued in extensive districts.

The long absence of rain had affected even the fish ; they decline the bait, and it was almost impossible to capture them. Slags and cormorants had come to the vicinity of the salmon ponds of Tasmania, and a few of the fish had been destroyed by them ; but a watch is kept in order that they may be shot as soon as they make their appearance. Commissioners were sent by the South Australian Government to examine into the state of the northermost runs in the district known as the Horse-shoe depression of early explorers, described by them as the partly dried-up bed of an inland sea, out of which neither man nor beast could pick a living, but into which since then squatters have found their way. The Commissioners went as far North as Yudanamutana, 250 miles from Port Agusta, and they report that nearly all the vegetation fit for pasture is gone. It is nearly two years since there was sufficient rain to saturate the soil. The commissioners are persuaded that the following estimate is rather below than above the facts : —235,152 sheep have perished out of 827,706 since the 30th of September '64, to the same date in '65, and 28,850 head of horned cattle out of 53,355. The horse stock has also suffered severely, 903 out of 2,145 being reported lost. Those losses do not include last year, (1864) increase of lambs and calves, for, with some trifling exceptions not worthy of notice, all have perished. Good seasons cannot be relied on in this district of country, and in dry seasons it is not calculated to carry so much stock as has been placed upon it."

This leads me to direct the reader to the declaration of the Prophet Joel, chapter i. verse 18. "How do the beasts groan, the herds of cattle are perplexed, because they have no pasture ; yea the flocks of sheep are made desolate." We will leave this branch of our subject, and proceed to the account of the Locust Plague.

⁋.The Thessally correspondent of the Levant *Herald*, writing from Larissa on the 16th December, '66, says :— We have escaped the cholera here, but we have incurred other calamities less terrible in appearance, but quite as disastrous to the population, and which have made a perfect desert of one of the most fertile districts of the empire. The locusts, whose destestable presence we had to endure last summer and autumn, have devoured all the wheat and a full half of tne other corn crops, except the maize, which fortunately has been spared, But for this the whole of Thessaly would have perished of famine. Wheat and maize form the greater part of the food of the people, and the failure of the other crops has raised maize to more than double its ordinary cost. Then, again, the locusts have destroyed the greater part of the tobacco crop, and what was left was destroyed by the peasants themselves in order to escape the tax, which they would have been utterly unable to pay." Says the same authority.—" I have just received from the captain of the Tirailleurs Algercius, not quartered at Budah, a letter entirely bearing on the plague of locusts, from which I extract the following paragraph : —'Our poor little village of Budah was thrown yesterday into a frightful state of consternation. Our splendid orange groves had hitherto escaped this horrible scourage. Four days since the first column of locusts took possession of our lovely gardens. Yesterday (July 20, '66) they arrived in so dense a cloud that the sun was darkened. In less than an hour the trees were literally covered by them. It is utterly impossible to convey an adequate idea of this plague. You see one day splendid gardens and luxuriant vegetation, the next day not a leaf or low blades of grass is left."

From the Battle Creek *Herald*, of June 1866 :—" The plague of locusts is causing sad devastation in Syria. In a recently received letter from Mrs. Bowen Thompson, dated Beyront, June 23, '66, it is stated that the accounts from Hisbaya are most distressing. It is worthy of remark that when the plague visited the country last year, the Christians exerted themselves to destroy the locusts and their eggs ; but the Mohammedans, who looked upon the locust as a great delivery, caught and ate them. The cholera has everywhere in the East followed in the train of the locusts, and the proportion of Mohammedans to Christians who have succumbed to the disease is beyond compare. A letter from a native of Hisbaga says that the locusts cover the whole land, and enter the houses as they have never done before. They have eaten up all the herbs, leaving nothing, and Hisbaya looks like a desert."

From the London *Free Press* of July, 1866 :—The Holy Land is again visited this year with a plague of locusts. A letter in a Manchester paper says :—' The valley of Urtas was first attacked and has now become a desolate wilderness. The olive yards of Bethleham, Beitjalah and Jerusalem were covered until the trees became a dull red color. They are now barked white. But yesterday (June 2,) will be a day long remembered. From early morning till near night the locusts passed over the city in countless hosts, as though all the swarms in the world were let loose, and the whirl of their wings was as the sound of chariots. At times they appeared in the air like some great snow-drift, obscuring the sun, and casting a shadow upon the earth. Men stood in the streets and looked up, and their faces gathered blackness. At intervals those which were tired or hungry descended on the little gardens in the city, and in an incredably short time all that was green disappeared. They ran up the walls, they sought every blade of grass or weed growing between the stones, and after eating to satisfy, they gathered in their ranks along the ground or on the tops of the houses. It is no marvel that as Pharaoh looked at them he called them *this death.* See Exodus the x, 4, 5, 6 verses. To-day (June 2) the locusts still continue their work of destruction. One locust has been found near Bethlehem measuring more than five inches in length; it is covered with a hard shell, and has a tail like a scorpion."

I am now done with this chapter of plagues ; and at least, it looks to the writer that they are one of the sure tokens of the near approach of Christ to Judgment.

The "Signs of the Times," your attention now call ;
Spurn not what is said as unworthy a thought,
Perchance you are wrong in some things you've been taught ;
If thus, it will do you no good so to be,
So come, let us " search," that the truth we may see,
For things of great moment, on this may depend,
As great and momentous, as those they portend ;
Then let us be sober, that truth may be known ;
Be that what it may, or by whom it may be shown.
The twenty-fourth chapter of Matthew you're told,
Has meaning, no mortal can ever unfold,
Which makes you conclude that the Lord will not come,
For many a year, to receive his "Bride" home.
We nowhere are taught, that the year'll be concealed,
Should the "day" and the "hour," be never revealed ;
So do not, my friend, with the "false prophet" say,
" My Lord, doth his coming, a long time delay.
The "fig tree" is budding, the "summer" is nigh,
God's herald's are sounding, the message be "nigh,"

And soon, he will "tarry" no longer for us ;
Though "tarried" he has, and the Scriptures say thus :
But soon, when our faith tried enough shall have been,
The wicked, have filled up their "measure of sin,"
And when the "five virgins," that helped swell the "cry,"
With zeal very ardent, and voice very high,
Have been "scoffed" from the field, and repose from the toil,
Till the "Bride-groom" shall tarry too long for their "oil" ;
Then, most surely he'll come, and the "Bride" will receive,
And the scoffer, and sleeper, on earth he will "leave ;"
It is then, that they'll "call," but the "harvest" is past,"
The "summer is ended," and their souls lost at last.

CHAPTER III.

ON CRIME.

Is this the nineteenth century ? Is this the eighteen hundred
and sixty-sixth anniversary of the inauguration of the Christian era, and of Him who was to bring " peace on earth and
good will toward men ?" This saith the Battle Creek
Herald, of 1866 :—" Has our boasted civilization and our
religion, after all, resulted in what we see around us ? Five-
tenths of the professed Christian world in open war, or,
breathlessly awaiting the dread moment when ' Red
Battle' shall stamp his foot? Have all our improvements
in locomotion, in transmission of intelligence, production
and exchange of commodities, only resulted in giving concentration, strength, and deadly effect to the worst and
most malignant passions and faults of our nature ? So it
would seem. We repeat it, nine-tenths of the so-called
Christian world are in arms, or may, at any moment, be en-
gaged in deadly strife. Not two years have elapsed since
the dominating nation of the Western World closed the
most terrific and bloody contest known in the annals of
mankind. How stand her neighbors on this continent now ?
Canada feverish and unsettled, after the first onset of a
subtle and persistent foe ; few years since we had neither
volunteers and but few soldiers. Truly, the question might
well be propounded, what is all this drilling and prepera-
tion for if not preparing for Armageddon ? Mexico, 'the
theatre of a prolonged and implacable struggle ; Chili, Peru,
Bolivia and Equador struggling against a monarchical
enemy ; Brazil, with the Argentine Republic, and Buanos
Ayres in an unequel contest against Paraguay ; Jamaica
still smelling slaughter ; the embers of revolution not yet

cold in Hayti and San Domingo. In Europe, Austria, with all her dependencies and her sympathizers in the Germanic Confederation, arrayed against Prussia and Italy; France silently arming in view of eventualities apparent to every mind; Russia and Turkey confronting each other in the principalities and threatening a conflict that must drag England into another war, which may re-light the fires of insurrection in India; Ireland under military rule, and heaving with revolution; Spain also in arms, and no one can say as yet, what the end will be. Are all these omens, these throes, these prospective and actual conflicts, portents of "Armageddon?" Can Christian, Philosopher, or Statesman reconcile all these conditions and fearful actualities with the theory or principles of religion and civilization? Are these terms only empty and cant phrases for bad men and charlatans to play and juggle with? Can this really be the nineteenth century? Is this the fruition of the gospel preach'd on the Mount? Or are we, in spite of its instructions and injunctions, savages as before, with only improved facilities for murder."

DIVORCE IN MASSACHUSETTS.—From the New York *Herald* :—" About 1,600 divorces have been decreed in Massachusetts in six years, of which 584 were for desertion, 553 for criminality, 132 for cruelty, and 42 from other causes. It is known that 1,316 were decreed in the five years that ended in May 1866. And at the same rate during the last eleven months, it may be assumed that the grand total is not far from 1,600."

From the Weekly *Globe.*—" Lives lost by the American Rebellion :—The War Department computes the number of deaths in the Union Armies, since the commencement of the war at 250,000, and of the Southern soldiers, at least 225,000, making in all 525,000 lives that have been lost to save a few Niggers."

> Six thousand years are nearly past,
> Since Adam from thy sight was cast;
> And ever since the fallen race,
> From age to age are void of grace.

From the New York *Tribune* :—" Crime has enormously increased in Washington since the reduction of the army. The police arrests last quarter numbered 9,122, an increase of over 3,000 beyond any previous fourth of a year. Plunder and robbery are the chief offences."

From the London *Times* :—" Out of 53,835 children born in Paris during 1865. 38,967 were legitimate, and 11,868 illegitimate. Talk of converting the far off Heathen." The writer of this book wants you first to commence at home. From the last authority :—"Curious Statistics.— Some strange statistics of matrimonial life•in Paris have just made their appearance. During the past twelve months of 1866, 2,344 wives have fled the conjugal roof without leaving their future address ; of husbands who have done likewise there are 4,427 ; of married couples legally separated (not divorced), there are 7,115 ; of ditto who have agreed to live apart, 5,340 ; of'husbands and wives living at daggers drawn, 31,912 ; of happy couples, 54 ; of mutually indifferent, 61,430. These facts are ascertained from the spies employed by the police, and the complete-ness of the espionage thus exercised is illustrated by a case stated in a Paris letter. A pastor had some doubts of an Eng-lish family lately settled in his parish, and who had borrow-ed $200 from him The pastor being on intimate terms with a *chef de devision* at the prefecture of police, stated the case. The chef inquired the name and address, rang a bell, desired his clerk to bring him Register C., and under ·that letter the proceedings of the family during two years' residence in the country parts of France were accurately recorded. Nought was set down in malice, but every fact connected with them carefully inserted in the register."

. From the Weekly *Globe* :—" There are 34 gaols in Up-per Canada and 20 in Lower Canada. Their inmates for ·'65 are reported as follows:

	U. C.	L. C.
Men, ...	3,962	3,571.
Women, ...	1,985	2,744.
Boys under 16, ...	311	257.
Girls do	103	83.
Total,	6,361	6,655.

This table shows a different result from that obtained by classifying the inmates of the Penitentiary in a similar way. Though Upper Canada sends by far the larger proportion of convicts to the Penitentiary, she does not send so many to the common gaols as does the less pupulous Province of

Lower Canada. The gaol of Montreal is credited with nearly two-thirds of the prisoners committed in all Lower Canada. No less than 1,938 men, 1,891 women, 156 boys, and 51 girls—total 4,036—were imprisoned at Montreal during '65. Of the 13,016 prisoners reported from all the gaols in both Provinces, only 7,255 were suffering their first imprisonment."

New York Daily *Herald*, Jan. 1860, says:—"Although prepared for an unusually extensive budget of criminal statistics, we confess that we are astonished, and startled at many of the revelations contained in the document. It appears that the number of arrests for offences of all grades, amounted to the enormous number of 68,873, or about 14,000 more than the previous year. Crimes of violence toward the person have increased in a still greater ratio, the total number being 995, against 620, in 1865. The report shows that there are 1,200 'daughters of perdition' in the concert saloons, and that there exists in this city and Brooklyn the fearful number of 10,000 places where intoxicating drinks are sold, over 8,000 of which are unlicensed. The License law and the Excise law are nearly inoperative. Bad and dangerous as the tenement houses are for habitation, it appears there are others which are worse. In one precinct there are 60 places, or dens, where the wretched poor, the criminals and depraved resort to lodge, paying from ten to fifteen cents per night for miserable accommodation. The places are chiefly in cellars, with naked stone or brick walls, damp and decayed floors, without beds or bedding fit for human beings. These dens are filthy beyond description, overflowing with vermin and infested by rats. In these hideous places are packed nightly an average of 600 persons—men, women and children—white and black sleeping promiscuously together, without regard to family relation, and exhibiting less of the impulses of decency than the brute creation. They consist of drunken wretches, male and female, beggars, rag-pickers of the poorest sort, sneak thieves, juvenile pimps, ragged and drunken prostitutes, and others of the same vile class. In the course of this revolting record of crime, vice and immortality, we observe the statement, that certain laws and ordinances are not enforced because the justices are dependent for their places upon the very offenders they are called upon to punish."

From the Battle Creek *Herald* of 1866:—"In Brooklyn, the 'city of churches,' there are over two thousand drinking shops, or about ten times as many as there are houses of worship. The cities contain 10,000 open haunts, in which rich men are made poor, sober men made drunkerds, happy families made wretched, and ruined. Two-thirds of all the rioting, harlotry, theft and pauperism of these twin cities issue directly from these dram-shops. During the war, many of them were nests of disloyalty. In each of these places—where death is dealt out by measure—there is an average daily expenditure of ten dollars for strong drink. This gives a total expenditure of $100,000 a day, of 700,000 a week, and of 35,000,000 of dollars a year This statement is so frightful that it seems incredible, yet it is rather under than above the actual facts. One of the Metropolitan Police Commissioners informs me that ten dollars a day, is a moderate estimate of the average sales of each dram-shop in the police district. Here, then, is more money expended in the purchase of liquid poison than is employed in sustaining all the churches, all the public schools, and all the charitable institutions of both the cities."

Judge Capron, of New York, in a recent charge to the Grand Jury, thus speaks of crime and drunkenness in that city:—" I would urgently solicit the particular attention of the Grand Jury to the law relating to the sale of intoxicating liquors. The retail of this article without license in this city, is a crime punishable by fine and imprisonment. Ten thousand dram-shops are at this moment occupied for that illegal traffic in this metropolis. The deluded victims of the illicit trade are reeling around us like spectres, to haunt the mercenary authors of their ruin, invoking from the cells of their gloomy prisons, imprecations loud and just against that partial administration of the law which consigned them to degradation and sorrow, but permit their destroyers to revel in luxurious pleasure, and even rewards them with official honors. During this year 342 convicts have been recorded in the Court of General Sessions of this city, and 4,215 in the Court of Special Sessions. In the same time there have been 138 acquittals in the former Court, and 694 in the latter. Of all these persons it is satisfactorily ascertained that only 94 were sober when the subject matter of the

3

accusations occurred, and all but 181 were habitually intemperate. More than one-half of this multitude were the victims of squalid poverty, and their crimes were induced by real want. 'Gentlemen, in view of these statistics let me inquire: How shall the wretched inebriates among us be reclaimed from their degradation and elevated to the dignity of men and women? How shall the humble but respectable poor, who are accustomed to drink, better their condition? Can it be done by all the efforts of persuasion and benevolence while dram-shops allure them on every hand? while the heartless proprietors are permitted with impunity to deal out poisonous mixtures under the false name of rum, brandy, or whiskey, and clutch their scanty wages, thus robbing them of food and raiment and shelter? No, the temporizing political economist may dissemble on this subject, the cold moralist may query, and the flash theorist may cavil, but nevertheless 'truth is mighty, and will prevail.' You can remove excessive dram-drinking only by removing the cause; to do that you must close the dram-shops, where the poison is sold. And in the present state of the law on this subject, to close those ante-chambers of prisons, alms-houses, and asylums, you must when complaints are properly before you indict the venders of those poisonous mixtures that craze the brain, and burn up the vitals of the drunkard. This is the only remedy for the evil that the law now affords, and this remedy does exist. I am aware of the fact, that since parts of the Prohibitory Law were decided to be unconstitutional by the mere majority vote of the Court of Appeals, but which are still held to be otherwise by a large number of the very soundest jurist of this State and nation, the suggestion that no law exists in this State against the free sale of intoxicating liquors has obtained advocates; but it is a clear proposition that no such consequence can possibly follow from that decision. There is a law on the subject of such sales, and the only sensible inquiry is: What is the law? This is not the appropriate place to submit an elaborated argument on this question, but I feel no reluctance to assume all the responsibilities of my position for this expression of opinion, that the unlicensed sale is now a misdemeanor."

There was hard times before, in the days of the flood,
When nothing was done but the shedding of blood;
When righteous old Noah, went into his boat;

And left all creation to sink, or to float, crying o' dear.
So up and be ready, for vengeance is near,
As God all the prayer, of his faithful will hear ;
And the few lurid mornings that dawn on us here,
Are enough for its woe, full enough for its cheer.

CONSUMPTION OF TOBACCO IN EUROPE.—In the City
of Hamburg, Germany, the manufacture of tobacco gives
employment to more than 10,000 persons, who turn out
160,000,000 cigars in a year, valued at $2,000,000. From
Havanah and Manila, Hamburg imports 18,000,000 cigars
a year, making an aggregate including its own production of 160,000,000 cigars, 153,000,000 of which are ex-
ported, leaving 15,000,000 for home consumption, allow-
ing 40,000 cigars a day to an adult male population of
45,000. In England, with a population of 21,000,000, in
1821, the consumption of tobacco was 15,598,152 pound,
an average of 12 ounces per head for the entire population ;
in 1831, with a population of 24,410,439, the consumption reached 19,583,841 pounds, or 13 ounces per head ;
and in 1841 with a population of 27,019,672, the consumption was 22,309,360 pounds, or 13½ ounces per head ; and
in 1851, with a population of 27,452,692, the consumption was 28,062,541 pounds, or 17 ounces of tobacco
per head, showing a steady increase. In France, the consumption is 18½ ounces per head. In Denmark, it is 70
ounces per head ; and in Turkey the consumption is even
greater. Enough of tobacco is smoked to keep every
poor man in the world well off, and yet, is smoked in the
air. Christians at least should not do this, or give their countenance to such a wicked crime.

How vain the delusion, that while you delay,
Your hearts may grow better, by staying away ;
Come wretched, come starving, come just as you be,
While streams of salvation, are flowing so free.

From the Weekly *Globe*, of 1866 :—" The unusual statement of crimes and disasters during the last six months
has been often remarked, and it seems the subject is now
about ripe for conclusions of figures. That the public have
noticed so large an increase in this fatal species of home
production may be sensibly attributed to the lapse of a
great war, and the revelation and recoil of the passions
which it absorbed. If some have been skeptical as to
the extent of individual outbreaks on society, it has
been for want of statistics, and partly because crime, which

has doubled so much in catastrophe of late, has been for-
gotten in other misfortunes, of which the last half year
has been full   Regular murders has almost found oblivion
in railway disasters.   But every chord of suffering life
seems to have been stricken, and we have heard of a host
of nondescript assaults upon the 'house of life,' beside
homicide, marricide, parricide, matricide, suicide, infanticide, familicide, patricide, attempted, hardly paus-
ing at cosmicide, which we interpret to be a violent
taking of life in honor of the journal which has made the
greatest display of murders.   Railway slaughter renders
necessary an addition to the vocabulary, and we have it to
hand, namely, viatricide, or murder of the traveller.   Altogether, here is an interesting field of enquiry for our
Kennedys and DeBows.   The entire estimate of the
capital crimes committed in the United States in the last
six months, it is doubtless impossible to give; but from
April to October, 96 murders are counted from metro-
politan files, exclusive of 12 manifest homicides and half
a dozen assassinations by indians and guerilles, twenty-
five murderous attempts are recorded, 9 cases of wife-
murder, and 7 of attempted wife-murder, which seems
to have been a terrible speciality with male criminals in
New York and New Jersey.   In contrast with this there
appears to have been but two husband murders.   Eight
instances of familicide, or murder of a family, including
three persons, make the most shocking feature of the
category.   Four fratricides, two parricides and matricides,
two double and two quadruple murders, eight infant murders, half of which were cases of abortion, were recorded.
In the West and South-west three Lynch Law trials took
place.   Singularly, amid all this excess of life taking, we
only find 88 suicides.   The great mass of murders and
crime generally were perpetrated between June and September, viz: 15 in June, 21 in July, 20 in August, and 26
in September.   The last month was by far the most fatal
with respect to railroad accidents, although it was supposed that had its climax in the previous summer months.
We conjecture that about 150 persons have lost their
lives by criminal hands, not remarking those who have
perished by the railway juggernaut.   Eccentricity and
monstrosity seem to have been prodigal in the chapter of
the criminal calendar just closed.   Of extraordinary cases
in America, most all of which seem to have been brute

outrages of a multiple kind, may be mentioned, the Joyce
tragedy in Roxbury, Mass.; the double murder at Sum-
merville, Pa.; the familicide in Tennessee; the wholesale
execution of Judge Wright and his four sons in Missouri,
and the Storkweather family murder. But the great
crimes of Europe have far exceeded all these in general
wonder and mystery, and it is seldom that a grouping of
such strange elements are found together in the social
history. In systematic wickedness and depth of motive,
they show a far older experience than American crime,—
intellect, science, a sort of philosophy, and even religion,
were sunk in the plot and women in the deed, with the
spirit of a young Lady Macbeth, a Fosco, a Thenardier.—
How Constance Kent, a child of 13, came to put away her
brother, is still a study; Dr. Pritchard, a medical reviewer,
daily resumed the patient task of killing his wife by slow
poison; Mrs. Winsor kept a hospital for infanticide;
Southey murdered the children of his mistress with a
parade of Malthusian philosophy, for fear that they would
starve, omitting, of course, to kill himself; and a Swedish
Priest poisoned a dozen of his parishioners with the sacra-
ment, out of pity for their wretchedness. Crime of this
complex character seems weird, appaling, and extrava-
gant beyond expression. Of the proportion of crimes in
general it is scarcely possible to arrive at conclusions;
but it should be remarked that twenty cases of nameless
outrage—a class of crime which seldom invites record—
were printed in the last six months. Curious and mon-
strous among other flagrancies were the placing of torpe-
does on a railroad, an attempt to throw a train off the
track; the burning of three houses by a girl fourteen
years old; four highway robberies by a boy of eleven
years of age; highway robbery by a politician; a mail
robbery by a post-master, and the malign biting off of
noses. Let the reader add to these the publication of ob-
scene books and papers, and the defacing of natural
scenery, several cases of which came under the law. An
ex-Congressman was also convicted of subornation of per-
jury. Large operations of robbery appear to have been
in proportion to other crime. The succession of several
robberies of banks within a short time, and the apparent
ease with which they have been robbed, have already
called forth words of caution from the public journals
against a state of financial insecurity. The following

statement of bank and kindred robberies alone will show
that this warning was not without reason of facts:—
January, '66, Bank of Crawford, Pa., $150,000; March,
National Central, N. Y., $50,000; Banking House, Bland &
Louisville, $50,000; May, Bank Walpole, N. H., $45,000;
June, Bank Wellington, O., $100,000; August, Banking
House, Portland, $25,000; September, Bank Concord,
$300,000; September, Treasury, Texas, $30,000; Sept.,
Adams' Express, $25,000; this comprehends only robber-
ies committed by outside parties, and we cannot pretend
to say the list is complete. A few of the robberies were
very remarkable, in the instance of the Crawford Bank
(evidently a worked-up robbery), it seems that while the
cashier was at work in the evening the thief entered, ex-
tinguished the gas, and made away with the bonds before
any light could be obtained. Mr. Bland, of Louisville,
was imprisoned in his own safe, and nearly suffocated to
death. The Concord Bank robbery is recent, and well
known. A list of Bank robberies would not be complete
without a statement of the immense defalcations and
swindles of the past six months, which have assailed the
safety of banks from within as burglars from without,—
we append a genuine list:—May. Bank New Haven
Savings, $100,000; August, Bank Phœnix, (Jenkins')
250,000; August, Banks, &c, New York, (Ketchum's)
$4,000,000; August, Erie Railroad bonds, (Jones') $500,-
000; August, Custom House, Memphis reported, $1,250,-
000; August, Quartermaster at Paducah, by a clerk, $25,-
000; September, Government bonds, Bliss, N. Y., $36,-
500; September, Revenue Collector, Ohio, defaulter, $90,-
000; September, Auerbach swindle, Louisville, $80,-
000; September, Railroad bond forgeries, (Gladwin) $204,-
000; September, by a New York book-keeper, $10,000;
October, Government bonds by General Bristow, Lynch-
burg, $80,000; this list does not include the alleged pay-
master frauds, navy-yard frauds, and official corruptions of
which the papers have been full It must not be omitted,
however, that a State Treasurer of Ohio, was removed
from office on charge of heavy embezzlements. Of important operations in the professional line may be mentioned the robbery of Mr. Veazie, at Albany, of $11,000; that
of $25,000 from an Indiana farmer; $48,000 at Chatta-
nooga; the $20,000 burglary at Detroit, and the $50,000
at Philadelphia. Forgeries have also been very numer-

ons and extensive, but it is, of course, impossible to give statistics; and, kind reader, if I was to enlarge on this chapter on crimes, as I might go from village to village, from town to town, city to city, country to country, lift up the screen that hides the enormous wickedness, and show at a glance the crimes of the earth, I should fain hope that you would not pervert your understanding, and say as a number of our professors do :—" That the world is getting better, and nearer conversion ; whereas, the truth is, that evil men and seducers shall wax worse and worse." See 1st Timothy, iv, 1 ; 2nd Timothy, iii, 13 ; Matthew, x, 30.

> Behold, on flying clouds he comes,
> His saints shall bless the day ;
> While they that pierc'd him sadly mourn,
> In anguish and dismay.
> Then haste thee, O haste thee, whilst yet 'tis 'to-day',
> We 'know' that the 'vision' cannot long delay ;
> Soon, Daniel will stand in 'his lot' with the blest,
> And you, if you're sav'd will be sav'd with the rest.
> Come my Lord, thy wright maintain,
> And take thy throne, and on it reign ;
> Then earth shall bloom again ;
> Oh come, come away
> Night soon will be over—
> And end'ess day appearing;
> Away from home ; no more we roam, O come, come away ;
> With sweetest notes of sympathy,
> We pray and praise in harmony,
> Love makes our unity, O come, come away.

I shall not trouble the indulgent reader any further with this chapter, but will turn to the chapter on Famine, which will not contain but a mere sketch of what has taken place the last three years; and you have my word for it, and watch it, for my serious conviction is, that the next three years will tell a sad tale of famine, pestilence and war, with crime, &c.

CHAPTER IV.

FAMINE.

The famine in British India; awful scenes of suffering and death. The London *Times* has the following from Calcutta, August 3, '66 :—" The mortality continues to be frightful; in four villages which an Englishman visited, there were not ten houses that did not contain one or more dead bodies, in another small place there were between four and five hundred dead, most of them unburied. The truth appears to be that the boasted administrative machinery of the Indian Government has completely broken down ; plenty of food has been bought, but somehow or other it does not reach the starving ; ships laden with rice are unable to discharge their cargoes for want of boats. Much of the misery is attributed to the fact that the authorities have given orders that only eight annas' worth of rice shall be sold to one person at a time. 'Every day,' writes a correspondent, 'there were hundreds of people coming and laying down their money, prostrating themselves on their faces and hands, begging to buy rice ; but the relieving officers cannot sell it, owing to the orders they have received.' In Calcutta alone, 200,-000 persons are subsisting on native charity, which has proved far more effective than the organization of the government. The news from the famine districts in Bengal continues to be very distressing, and was beginning to make itself felt in the streets of Calcutta, It is stated that crowds of sufferers from the Nudda division, where the failure of the crops has deprived them of food, are finding their way into the metropolis in the hope of obtaining assistance and sustenance, and the streets now present very distressing pictures of suffering humanity. Fathers and mothers in a dreadful state of debility are selling their emaciated offspring to passers by for 3 to 4 shillings each, and are seen searching for a few grains of food among the offal cast out at our doors. The rind and stones of mangoes are eagerly caught up and sucked, in the vain hope of sustaining life a little longer thereby, and such refuse as a dog would reject is eagerly devoured. Crime has of course increased in consequence of this influx of men driven to desperation by hunger, and there is a melancholy crop of thefts and burglaries. The famine is very sore in Orrissa ; in the fifteen districts affected, but chiefly in the three districts of Orrissa, and the adjoining country of Midnapore, 75,000 are daily fed by public charity; if you double that for the numbers fed privately, and chiefly by Hindoos, you will be still within the truth. Out of Orrissa and Midnapore half of the destitute are professional beggars; in these provinces nearly all are the laboring poor, and the lower class of agriculturists. The largest number of deaths from starvation in Orrissa and Midna-pore reported in one week, is 3,500, and in the Southal country some less. The average number of deaths report-ed to the authorities during the past six weeks in those districts, is not less than 2,500 a week ; add to these the deaths witnessed by no human eye in the far interior, where aid is never penetrated, and you will have by no means an exaggerated idea of the state of Orrissa and Mid-napore. Says the writer, I will not harrow the feelings of your readers by the details of cases which appear in the daily papers here, reported by eye witnesses of the jackals eating the corpse of one wretch while they wait for his companion who is dying, or of the child taken from the breast of its mother, who has been dead two days. We know still less of the state of Ganggam, the Madras district immediately to the South of Orrissa. The Famine began in October last, it became so grievious by December, that gold, silver and brass-work sold at twenty per cent below the usual rates, and the magistrate of Pooree urged the establishment of a relief fund and public works as well as the revival of the salt manufacture. The December crop was saved by the rain, but it was so scanty, and the peas-antry had to give so much of it to the landlords and money-lenders in repayment of advances, that by the middle of February prices again fell to the level of starva-tion, rice got to one shilling for five pounds. The people managed to struggle on, till by the beginning of April they had exhausted their stores, and from the first week of that month, when the missionaries and the magistrates appealed for public assistance, the famine in Orrissa and Gangam dates. The public began to subscribe, when it was remembered that there was an unspent balance of $312,500 of the North-west Famine Fund. Government at once gave up that sum to the Board of Revenue, which was guilty of the incredible folly of informing the public that no more subscriptions would be necessary. The Chamber of Commerce informed the Bengal Government that they wished to form a great central committee, but they were politely snubbed. The necessity for importing rice into Orrissa was urged ; for a month the board refused to see the necessity of interfering with private trade. Where, all this time was the local authority, the Lieuten-ant Governor ? In the hills of Dargeerling, with all the neads of departments, except the board from which he was separated by a three days' post. Sir Cecil Beadon had himself been in Orrissa a few months before, when he told the people the land tax would be raised at the new assessment; and of course he had not seen any famine then. Why should he be disturbed in his cool retreat? The Governor General ordered him down to the post of duty, and with the first showers of rain he entered Calcutta.— He has since presided weekly at the meeting of the board, with which our native Zemindar and one English merchant have been associated. Rice has been sent down in large quantities, the government of India having advanced $1000,000 for the purpose; but still the board says no subscriptions are wanted, when 2,500 are dying every week. On a surf-beaten coast in the height of the monsoon it is difficult to unload rice, and much has been lost; but no supply of labor has been sent to carry the rice into the interior for the dying, and to-day's telegram reports the price at Cuttack, on the 31st of July, '66, as still under five pounds for a shilling, while the showers are so partial as to be insufficient for the crops. What must be the state of things outside the country town, and away from the few relief stations? Not only so, but although it is evident to all, that there are no traders' stores of rice in Orrissa; much of the government rice is sold at the market rate quoted above, and one of the relief committee was ordered last week to sell the unpopular Burma rice at the high price of 8 pound for a shilling. Set against the calculation of 2,500 reported deaths per week, the facts that up to the end of July, after four months of severe famine, the board of Revenue has spent in the fifteen afflicted districts of Bengal only, $77,550 in grants of cash to relief committees, and $337,605 in the shipment of grain, while it is liable for $200,000 more in orders not yet executed, and you will be able to judge if there is not ground for at least the suspicion that, from want of forethought or foresight, bad management from inability to organize a system of relief equal to the vast area of suffering, the lower classes of Orrissa and Midnapore are perishing.— The misery will certainly not lessen in Eastern India till the end of October, and not then if more rain does not fall, so as to secure the great autumn crop, the failure of which last year is the cause of all the suffering. To the brief history I have given of the sufferings in India, let the reader reflect ; the suffering a few years since in Scotland and Ireland, as also, in South America, in London, Canada, and last year in Nova Scotia, with the Red River affliction just now going on. Listen to what the Bishop of Rupert's Land says on the want in that region :—" The Committee here are very thankful for the great kindness which has been shown towards the poor sufferers in this country by many citizens of St. Paul and Milwaukee.— It is very much to be regretted that such a report received currency, as you refer to, respecting the distress being exaggerated ; on the contrary, it now threatens to be severer than, I suppose, any apprehended. The poor people of this country have been so accustomed to shift for themselves in times of difficulty, that few at all realized the state of things. But the total failure of the buffalo hunt, partial fall, and the scarcity of rabbits, combined to produce the danger of most serious distress, and though a great deal has been contributed, we find that the expense of freight will swallow up a very large portion of the gifts. There will also be, I fear, a very serious deficiency of seed grain for the coming season. There cannot be such fear of exaggeration of distress, when in an isolated region like this, the whole of the crops of every kind is absolutely swept away." Famine, therefore, I say is one of the sure tokens of the near approach of Christ ; mark what I say, if I am on the right track, the next three years will double the affliction of the last three years, but none of the wicked shall understand, but just like the days of Noah and Lot, none caring or desiring to be looking or loving the appearing of Christ, hence, on such he will come like a thief.

Light is beaming, day is coming,
   Let us sound aloud the cry ;
We behold the day-star rising
   Pure and bright in yonder sky.
         Saints rejoice now—
Your redemption draweth nigh.

We have found the chart and compass,
   And are sure the land is near ;
Onward, onward we are hasting,
   Soon the Saviour will appear,
         O, be cheerful—
Let the word your spirits cheer.

Hark, hark, hear the blest tidings,

Soon, Soon Jesus will come,
Robed, robed in honor and glory,
To gather his ransomed ones home,
Yes, yes, O yes—
To gather his ransomed ones home.

I shall now pass this chapter and leave with the reader to say what is truth. Our next chapter will be on floods.

CHAPTER V.

FLOODS.

From the weekly *Globe*. Sept., '66 :—" Cincinnati, Sept. 2nd. · Heavy storms visited this region yesterday and last night, causing floods that have resulted in seriously damaging provisions to the extent of hundreds of thousands of dollars. Dear Creek, a stream running through the eastern part of the city, was a scene of turbulent and destructive inundation. This morning before daylight, about 30 tenement houses were swept away. Longworth's celebrated wine cellars, containing 150,000 gallons, were overflowed, raising the temperature and causing the champagne bottles to explode in a deafening fusilade, and hundreds of casks of wine were afloat for a while. Three extensive tanneries were completely gutted by the flood, inflicting immense loss. The total damage will reach $400,000. Several persons were drowned, and a number of railroad bridges are reported swept away on the Dayton and Michigan and Maridta Road."

GREAT FLOOD IN THE NORTH WEST.—(Special to the New York *Herald*.)—" Chicago, August 14th, 1866. Complete returns have been received here of the condition of the wheat crop in the North West, since the storm of Saturday and Sunday, which was the severest of the season. The wheat in the northern tier of counties in this State has been materially injured. In the southern part of the State the crop has been pretty generally secured ; but in the north much of the wheat was in the shock, and has commenced growing badly. Along the Galena branch of the Northwestern Road, reports about the wheat are very unfavorable. The crop in Wisconsin is almost entirely ruined. The harvest there is two weeks later than in Illinois, and the floods of rain caught the farmers in the field. From one end of the State to the other come tidings of floods, inundations, bridges swept off, and the whole country for miles under water. The crops, both cut and uncut, are completely submerged, and are heating and sprouting to an extent that must ruin them. The loss of this crop must and will be severely felt all through the West. The crops in Minnesota are better off, although seriously damaged.— Ohio, Indiana, Michigan and parts of Illinois, Iowa and Missouri have secured their crops, and the quality is reported very good. In Minnesota the storm was very severe. At Rochester several rods of the Winona and St. Peter Railroad bridges were washed away. Four or five houses were swept away by the rise in the Zumbro river, which rose fourteen feet during the night ; so sudden was the rise in the river that the occupants of houses on the low lands, had scarcely time to escape in their clothes and save their lives, and on Tuesday morning eleven persons, men, women and children were rescued from trees, where they had taken refuge during the night. All the bridges between Rochester and Owatonna are swept away, and the railroad bridge near Casson is also gone. Near Houston, about twenty miles west of La Crosse, thirty persons were drowned by a sudden rise in the Root River, Minnesota ; twelve bodies have been recovered and buried. The Southern Minnesota or Root River Railroad is in some portions seven or eight feet under water. The town of Houstan is almost entirely submerged, as is also a considerable portion of the village of Rushford. Entire farms have been swept over by the raging current, the water even reaching above the top of the growing corn crops. The destruction of property is great, reaching, probably, to $100,000, or more. Such a flood was not known in this region before, although it is remarkable for the number and extent of its freshets. The heavy rains in Northern and Southern Iowa has caused very high water in the Cedar River ; bridges have been swept away, and much damage done. The bridge at Cedar Falls has gone down stream."

From the New York *Tribune*, Oct., '66.—" Baltimore, Oct. 14th. The rain has continued to-day almost without intermission, and is still falling. A heavy easterly wind prevails. All accounts agree that the quantity of water which fell in this region last night and this morning, was most extraordinary. A dumber of dams and bridges on the Rotapsco were destroyed. The extensive dam at Elapsville was swept off, and coming against the turnpike bridge at Elliott's destroyed it, The wrecks finally accumulated at a heavy stone bridge at Illchister, on the Baltimore and Ohio Road, 13 miles from this city, which gave way to the tremendous pressure. This bridge was considered the strongest on the road, and stood out against all former floods. A family of six persons and another of three were drowned ; several bodies were found to-day."

From the Weekly *Globe*, '37.—" The inundations in France were subsiding. Great damage has been done by the floods The Emperor had headed a subscription with 100,000 francs."

From the New York *Herald*, October, '68.—" San Francisco, Oct. 2nd. One of the most destructive rain storms ever known in AncArizona, commenced on the 7th of September, and lasted for three days, completely flooding the country. Many of the villages of the Rimo and Mancopa Indianson Gala River were entirely destroyed, large crops were swept away and Huppei & Co's. steam flouring mills at Runo village were destroyed, and a number of cattle belonging to Texas emigrants were drowned in the flood. Gov. McCormick and party, en route from Prescott, were caught in the storm and compelled to swim their animals to reach a place of safety. Telegrams from the interior report great damage done to the crops by the rains of two days past. The ship brought on from Glasgow reports having experienced a heavy shock of earthquake at sea on the 11th and 18th of Sept."

THE OVERFLOW OF THE NILE AND ITS CONSEQUENCES. —The apprehensions created by the rapid rise of the waters of the Nile are, unhappily, being justified by events. We have arrived at that period when a falling of the water should be noted, but, on the contrary, there has been lately a continued and considerable rise. The houses on the banks of the river in the suburbs of Boulak and Old Cairo, have been invaded by the waters and some portions of the quays have been destroyed. The Egyptian government is displaying most praise-worthy activity in meeting the threatened danger. It has employed all the means in its power to strengthen the dykes and to repair the banks. A careful system of supervision has been

established, and vessels laden with stones and materials
are stationed at short distances, ready for being towed to
any menanced point by the tug steamers on the river.—
Disasters are mentioned as having occured in Upper
Egypt, where the waters have swept away the corn crop
heaps in the fields bordering on the river."

Cario, Sept. 26.—Correspondence of the Paris *Moniteur*.
"The catastrophe at Bezandun. A meeting was held in
the upper room of Queen Street Hall yesterday afternoon,
for the purpose of hearing a statement from Dr. Muston,
respecting the catastrophe by which the village of Bezandun, situated in the French part of Vandois, has been destroyed, 1. Bonar presided. Dr. Muston, who addressed
the audience in French, stated that Farel, the companion
of Calvin, had labored in the country in which Bezandun
is situated, and that it has been the scene of trials and
martyrdoms for the truth's sake, both before and after the
Reformation. Bezandun was built on the steep slope of a
hill, consisting of a great rock covered with earth, so that
the village was supported on a stratum of soil lying on
the smooth surface of a slanting rock ; the houses were
built across the face of the hill, above these were the
gardens and vineyards of the villagers, and farther up,
the church ; down the hill streams used to flow from
springs at the summit of the hill, and this year, in the
month of May, when the rains were falling so heavily
throughout the south of France, these streams, instead of
increasing, as might have been expected, diminished, and
it was found that the fountains were nearly dry. These
symptoms of an approaching catastrophe caused no great
surprise at the time, but they were remembered afterwards. On the 31st of May, as a colporteur was passing
by, his attention was attracted by some strange sounds at
the church, and a few minutes afterwards the population
were greatly surprised by hearing three or four strokes of
the church bell. Immediately afterwards, in the houses
of the village, the windows commenced to break, and by
a subterranean movement the walls were displaced, so
that doors which were shut could not be opened, and doors
which were open could not be shut ; the people rushed
out of their houses, mothers dragging their children after
them, while tiles, chimneys, and the lighter parts of the
houses fell. They were no sooner out of their houses than
the roof of the church fell in, destroying the interior. It

appeared that the springs of water which used to flow
over the surface from the summit of the hill had, at the
same time, flowed underground, reaching at last the sub-
stratum of rock, which, being impenetrable, the water
collected till at last there was such a quantity that the
whole soil beneath the village became detached from the
rock and began to slip, carrying the village away with it
in its downward progress. It slipped, not all at once, but
at irregular intervals during twelve days, which enabled
the villagers to escape with their lives, although they lost
all their property. The village had entirely disappeared,
and the inhabitants were entirely destitute ; and it was
in these circumstances that he had been sent to friends
and fellow-Protestants at a distance, to request their aid to
enable the villagers to re-build their church and schools.
The whole pecuniary loss estimated at 65,000 francs or
$13,000. Mr. Robertson of the Gray friars and Mr.
Montgomery, of Innerleithen, commended the cause to
the liberality of Christians in this country, and the latter
stated that Dr Muston had been banished from Piedmont
twenty years ago, for having published a historical work
which had given offence to the Romish priests at a time
when their will was supreme in Sardinia."

Edinburgh *Witness.*—A correspondent of the London
Times gives a detailed account of the recent disastrous
floods in Northern Italy. Sept., 1868. He says :—" Over
the greater part of Northern Italy rain had fallen almost
without intermission, for twenty or thirty days, though
not in the Alpine district so as to cause any serious catastrophe ; but on the night of Sunday, the first day of the
week, the 27th of September, a terrible storm broke upon
the southern slopes of the Alps, producing, in a few hours,
vast torrents and avalanches, under which roads, houses,
and villages were swept away, and large tracts of fertile
country were sunk in mud and water. The storm of the
27th, however, proved to be only the signal of disaster to
crime. For a week since, night and day, there has been
but one continuous storm and cataract of rain, the results
of which are not yet known, but which has already
covered the whole valley of the Po with a series of inundations. On the morning of the 29th we left Luccrene to
cross the St. Gothard road in splendid weather and in total
ignorance of the catastrophe of the 27th. Immediately

after leaving Faido, the signs of destruction began, the
road had been torn up at intervals by torrents descend-
ing from the precipices above, and swept by avalanches of
earth, stone and timber. As we passed on, the destruction became worse, orchards, woods, vineyards, and
chalets were seen to have been hurled in a mass across
the valley, which they covered with ruins, and for long
tracts, not only had every trace of road disappeared but
every trace of cultivation itself, also. So that what used
to be once a rich country, teeming with produce, and
traversed by massive causeway, had returned to its primitive state of torrent bed, and primitive rock. At Bodio
the disaster has been greatest. The whole village was
swept by a torrent of mud and stone, which scarcely left a
house standing, and buried about twenty persons in the
ruin. The destruction was almost instantaneous. The
torrent, which descends from the mountain above, had
burst its channel and partially flooded the houses, when
about midnight on the 27th, a crash high up the
precipices was heard, and soon a stream of mud and rock
swept over the village and almost buried it out of sight.
As we reached Biaska, where the Breno joins the Ticino,
further disasters appeared. The Breno was rising more
violently even than the Ticino, threatened to cut the communication up the main valley, which a few hours after we
forded, did actually occur. The villages of this lateral
valley had been swept by avalanches, and in all of them
property and cattle, and in some, many lives had been lost.
In a word, the whole valley of the Ticino, which every
tourist will remember as a scene of continual beauty and
richness, has been desolated; for twenty or thirty miles,
its entire sources of industry have been destroyed, and
great tracts of it have been changed from the most fertile
soil into a mere desert of sand and rock. But what was
happening in the valley of the Ticino was only a specimen
of what was befalling many a valley of the Alps. At Bellin-
zona rumors more or less distinct were rife of singular disasters in all parts of the range. The St. Southard road, as
a great highway, is totally broken up on the southern side,
and will not be completely restored for months. The
Bernardina and the Splugen are also broken, and great
bridges destroyed. The Simplon road is for many leagues
fathoms deep in water. Before reaching Magadino, the
plains were seen for miles under water, and at length the

road itself was submerged. Here, with no small difficulty, and at some risk, boats were procured, and in the midst of a furious storm of wind and rain and lightening, the village of Magadino was reached, half sunk in the flood. The pier and all the offices at the wharf were scarcely visible, and the lake appeared to stretch right across the valley almost up to Belinzona. Towards evening the steamboat proceeded on her voyage down the Lago Maggiore. As each town on the lake was passed, it was seen that it was half sunk in the water. Locarno, Cannobio, and Luma showed only the upper stories above the waves; the road was itself submerged. villas, churches and towns in the midst of the lake. At Jntra and Palianza the greatest injury occurred. There the streets and houses were covered by twenty feet of water, and as they were exposed to the gale from the South, and the bay was choked with fragments of wood, several houses had been beaten down altogether, and many lives lost. On reaching the bay where the Tosa falls into the lake, it was seen that the whole Simplon road from Arona to Ornavaseo was completely under water, and indeed, the lower valley of the Tota, like that of the Ticino, was a simple arm of the lake. The great hotels and the villas with which this part of the lake was bordered, were submerged to their first and even second stories, the postal and telegraphic communication was cut off, the railway station at Arona was almost covered, and the granite posts for the electric wires just showed their tops above the water. Every town was more or less covered, and the inhabitants were hasting in boats to places of safety, and removing parts of the furniture and goods by ladders from the upper windows. The Ticino was unable to carry off the pressure of waters, and had flooded its whole valley for leagues down the Lombard plain. The Laggo Maggiore, which had risen about twenty feet, was still rising at the rate of four or five feet in few hours, and there was every prospect of a still greater rise. Nothing of the kind has been known in the memory of man, and the only tradition of such a flood appears to date from one hundred and sixty years ago."

    1.   O hail, happy day, that speaks our trials ended,
        Our Lord will come to take us home ;
             O hail, happy day—
        No more by doubts or fears distressed,

We soon shall reach our promised rest,
And there be forever blest ; O hail, happy day.

2. Swell loud the glad note, our bondage soon is over,
The jubilee proclaims us free;
O hail, happy day—
The day that brings a sweet release,
That crowns our Jesus prince of peace,
And bids our sorrows cease ; O hail, happy day.

3. O hail, happy day, that ends our tears and sorrows ;
That brings us joy without alloy ;
O hail, happy day—
There peace shall wave her sceptre high,
And love's fair banner greet the eye,
Proclaiming victory ; O hail, happy day.

4. We hail thy bright beams, O morn of Zion's glory,
Thy blessed light breaks on our sight ;
O hail, happy day—
Fair Beulah's fields before us rise,
And sweetly burst upon our eyes,
The joys of paradise ; O hail, happy day.

5. Thrice hail, happy day, when earth shall smile in gladness,
And Eden bloom o'er nature's tomb ;
O hail, happy day—
Where life's pellucid waters glide,
Safe by our dear Redeemer's side.
Forever we'll abide ; O hail, happy day.

Reader, I am done with the foregoing chapter, and ask of you, an earnest perusal of the sixth chapter, which is on hurricanes.

CHAPTER VI.

HURRICANES.

DISASTERS ON THE LAKES.—The Detroit *Free Press* publishes a long list of lake disasters during 1868. The number is 341—more than any previous year. Two hundred and fifty-five vessels have been ashore, as follows :— "On Lake Michigan, 107; Lake Huron, 50; Lake Erie, 65; Lake Ontario, 27; and Lake Superior 6. 89 total losses have taken place, viz:—Lake Erie, 24 ; Lake Huron, 13; Lake Michigan 34; Lake Ontario, 11; and Lake Superior, 2. The disasters which have occurred in the straits, far surpass those of any former year, and have been credited to either Lake Huron or Lake Michigan, or to which ever locality they happened in closest proximity."

From the Weekly *Globe*, Oct., 1868.—"A gale of unusual violence swept over Prince Edward Island and the north shores of the neighboring Provinces on Monday night last. A number of vessels were blown ashore, and some lives lost. Several buildings were also blown down. Loss about $10,000."

From the Weekly *Globe* of 1866.—"A fearful tornado. Galveston, Texas, July 17th. The steamer Harlan brings accounts of a tornado lasting three days, beginning at Indianola, on the 13th, and ending on the 15th. Four vessels were totally wrecked; of two of them there was not a vestige left. The steamer Patmos, which was anchored outside the bar, has disappeared; it is supposed that she foundered at her anchorage, and that all who were on board in charge of her, were lost; her passengers had previously been landed. No such storm has occured on this coast within the memory of the oldest inhabitants."

From the New York *Tribune*, July, 1867.—"A tornado. A terrible tornado passed over the village of Newbern, Georgia, on the 24th ult., levelling houses, fences, trees, &c., sweeping in an instant, everything before it, and killing as well as injuring a number of persons."

From the New York *Times* of Dec., '66.—"Shipwreck and loss of one hundred lives. We deeply regret to announce the loss of the Dutch barque Mercurius, Captain Smith, of 439 tons, bound from Amoy to Singapore, which melancholy disaster took place on the north coast of Bintang, and 100 Chinese passengers perished, as also the third mate. The Captain, in a letter dated Rhio, 16th instant, notices that the weather was very thick, and the vessel was driven on shore, having parted from two anchors."

From the Orono *Sun*, 1866.—"On Monday afternoon this neighborhood was visited by a terrific thunderstorm. The lightening was very vivid, while the peals of thunder were such as are rarely heard even in this climate. Torrents of rain poured down, which in some places did much damage to the now nearly ripe grain. At Port Newcastle a boy was instantly killed by the fluid. He was sitting by the stove in company with his mother, when he was struck dead. The bereaved parent escaped unhurt. At Enniskillen, rumor says two persons were also killed. At Bowmanville a woman was struck by the fluid, though not then killed. And with the wind and rain together much damage was done to the crops."

From the Weekly *Globe* of June, 1866 —" On Sunday a terrible hurricane visited Niedet. Barns were blown down and roofs of houses carried three or four acres, trees were uprooted, and places were burned by lightening Great damage by lightening. A heavy thunder storm, accompanied by torrents of rain, occured on Wednesday afternoon. Several places were struck by lightening in the vicinity of a block of twelve unfinished buildings on Warren street, Brooklyn. The block was nearly demolished by the lightening and the gale. One or two vessels were also struck by lightening St. John's Chapel was set on fire in this city, and the steeple in St. Teresa Church was struck."

From the Hamilton *Spectator*, October, 1866.—" Marine disasters. During the month of September, there were lost by fire, wreck, collision and other disasters, twenty-two vessels belonging to the United States, valued at $1,618,-000. There were three ships, one steamer, nine barques, three brigs and six schooners destroyed during the month. Thus far in 1866, there have been three hundred and eighty-nine vessels, valued at $19,682,800, lost by disasters at sea."

From the Battle Creek *Herald* of 1866.—" Every day some new calamity is reported of a most heart-rending character. We have had of late the great fire at Wiseasset, Me.; the destruction of 2,500 dwellings, beside many public buildings in Quebec, leaving 18,000 persons houseless; the loss of the Evening Star, on its way from New York to New Orleans, in which nearly 300 perished; terrible hurricane at the Bahamas; loss of a vast number of vessels at sea by the late gales, with hundreds of lives lost; the great floods at the West and South, causing the loss of millions of dollars in property; the devastation of the cholera in nearly all our cities; the great famine in India, with wars and rumors of wars, all over the world. Revolution follows on the heel of revolution, and all are working for the struggle."

From the Brighton *Flag*, 1866.—Storm in Prince Edward. On Tuesday evening last, 16th inst., a very destructive hail storm passed over the Township of Hillier and visited the village of Wellington, effecting much damage to property. The crops along the lake shore, from Consecon to Wellington, were almost destroyed; and such was the injury to the windows in Wellington, that some of the citizens of that place visited Brighton yesterday, procuring a supply of glass to repair the damage."

From the Banff *Journal* of July, 1866.—The Parish of Insh, in Badenoch, was visited with a thunderstorm, accompanied by hail and rain. The road from the bridge of Feshie, along the south of Spey, is rendered totally impassable, four bridges and several culverts having been carried away. On the estates of Inneerveshie and Nude much damage has been done to the growing crop, the soil in some places being wholly carried away, and in others buried several feet below the gravel carried down by the torrent. From Nude about forty sheep were carried away by the flood, and lost in the Spey. The Mansion house and offices at Gordonhall narrowly escaped being carried off by the burn which passes there. The hailstones were generally the size of a musket ball, and they were seen lying in the sheltered places unmelted for one week after they fell."

From the Weekly *Leader*, June, 1866.—Destructive hail storm. The township of Camden, East Sheffield and Richmond were visited on the 20th June, with a storm of wind, rain and hail that unroofed houses, blew down fences, tore up trees, and destroyed and battered down fruit, crops and everything in its course, peas, barley, wheat, and other crops were beaten as with a flail. Hail measured from one and a half to two inches in diameter, and when the storm abated were about nine inches deep on the ground. The storm left drifts of this shower of ice in some places from two to three feet deep."

From the New York *Times*, '66.—Disasters at Sea. Within the last week northern latitudes have been visited by a terrific gale, which have caused many shipwrecks and much loss of life. The cyclone seems to have been of wide area, and lasted for some time. On the 1st, we read that the ship Sebastopol was struck by it in latitude 26 deg., 39 m., longitude 79 deg., 38 m., and became dis-

masted and water-logged, the crew having to take to their
boats, and were rescued by a passing vessel. On the
4th, the British steamer Queen Victoria encountered the
hurricane in latitude 38 deg., 3 m., longitude 70 deg., 30
m., and foundered at sea, the crew and passengers taking to their boats and being ultimately picked up by an in-
ward bound vessel, and brought to New York. There are
many other casualties reported, but the loss of the Evening
Star is sickening. The late gale ought to act as a warning to owners of vessels, to provide everything which
will tend to the safety of them."

From the Detroit *News*, July, 1866.—"The hail storm in
Northern Illinois last week was very destructive, and the
hailstones in size were beyond predecent. One was picked up in Lanark three and a half inches in circumference
and over one inch thick. A peck was gathered up of
nearly that size in Lanark. Over four thousand lights of
glass were broken out at Blackberry Station. Hardly a
whole light of glass was left in town. In Elgin several
thousand lights of glass were smashed ; the hailstones
being seven inches in circumference. In the track of the
storm, which was about a mile wide, corn and oats were
completely cut down, and garden vegetables destroyed."

From the New Haven *Palladium*, July, 1866.—"Dreadful storm in Connecticut, and inflicted great damage.—
Out in the country the damage was immense, especially to
the crops. In Orange, the corn, oats, grass, &c., are almost
wholly prostrated. One of the stacks of hay on the
meadow was struck by lightening during the storm and
took fire and, notwithstand the violent rain, was burnt to
the ground. The storm was more violent in the towns to
northward than here. In North Haven the electric discharges were fearfully frequent. Near the residence of
Mr. I. H. Thorp, the lightening struck ten times on ten
different trees within a quarter of a mile ; the trees were
of different kinds. We do not remember that we have
ever heard of a more remarkable frequency of electrical
discharges within so short a distance. All along the line
of the Canal Railroad we hear of barns and houses prostrated by the wind, and of buildings struck, while the
crops suffered terribly, and the country looked like desolation indeed. At West Cheshire we hear that several
buildings were unroofed. In Meriden the hail storm was

exceedingly violent, and the thermometer suddenly fell from 960 to 710. Randolph Lindsay's grapery was entirely destroyed by the storm. The steeple of the Hanover Congregational Church was broken off by the violence of the wind, about thirty feet from the top. The broken part tumbled over and fell striking on the point and sticking deep into the ground. The drying shed of the American Comb Company, at their bleach works in Hanover, was completely wrecked by the storm. The glass roof, 200 feet long, was entirely destroyed."

From the London *Free Press*, August, 1866.—" The thunder storm of last week has done great injury to the cereal crops by laying them down as far as the crops extended, A great many corn, fields have been cut and partially housed. On Wednesday last the thunder was accompanied by a hail storm in several places of a character more severe than was ever remembered. The town of Windsor suffered the most, scarcely a window in any of the building in one aspect escaped destruction; several large trees were shivered to pieces in various parts, and several persons were killed by the lightening."

From the Montreal *Gazette*, July, 1866.—"Hail-storm.— A hail storm passed over the parish of Reputigny on the 11th inst., which entirely destroyed a portion of the crops and killed many cattle. The hail-stones were extraordinarily large, and so completely covered the earth that a sleigh could have been fitly run over them."

From the Brighton *Flag*, 1866.—" Tornado. We were visited on Monday last with a severe thunder storm, accompanied by a terrific gale of wind. In the village it did considerable damage by breaking down trees and destroying gardens by the hail. Among other disasters a new building belonging to Mr. H. C. Betters, was blown down with a terrific crash, its timbers are literally broken to pieces. The joiners were working in the building when they saw the gale coming, they fled, and had not left it more than three minutes before the building was lifted from its foundation and dashed to the ground a heap of ruins. We also learn that the storm passed through the centre of this township, doing great damage to the crops, fences, &c."

From the New York *Herald* of October, 1866.—" Terrible storm at Nassau. Havana, Oct. 19th. A terrible hurricane commenced in the Bahamas on the 30th ultimo, and lasted two days. Almost half of the town of Nassau was destroyed by the storm. Houses were blown down, roofs carried away and trees uprooted. Trinity church was demolished, the government house lost part of its roof, and the roof of the Marine Hospital was entirely blown off. Vessels were driven ashore and knocked to pieces, and wharves were demolished. The neighboring islands suffered in the same degree, and a large number of vessels have been lost or damaged. This hurricane is the severest in the rememberance of man. A correspondent writing from Newbern, Georgia, 1867, gives the particulars of a tornado of unparalleled fury. One woman was blown a distance of 400 yards ; her house was found 100 yards distant from its foundation, fragments of which fell six miles distant. The writer says :—The whole compass of its visible ravages is comprised within an area not exceeding 200 yards in width, the track of its chief violence is even narrower. Scarcely a tree is left standing where it passed, of the fallen ones some lie at every point of the compass. Every out-building through the entire course of the tornado from the point of first attack, beginning with the barn and stables of Mr. Smith, was swept away, some of them to their very foundations.— The same is true of fencings. In one case where there was a lane the rails were heaped promiscuously between the two. Of the four dwellings occupied by white families, only one, that of Mr. Joseph Kinney, was left standing, roof broken in and shattered. Outside of the main channel the houses of Mr. Beeland was damaged by flying timbers. The buildings' occupied by Mrs. Moss was carried away to the floor, which was literally covered with the debris of chimneys and the tops of fallen trees; yet of the seven in the house at the time, all escaped with life save one. The kitchen was occupied by Burrel Binford (colored) and his wife and another woman. Of these, Burrel was blown to a distance of fifty yards, and killed, his wife was severely injured ; she was carried twice in the air, and says, she saw fragments of timber flying as thick as leaves in an autumnal gale. The dwelling of Dr. James H. Montgomery was lifted from its foundation, turned over and dashed to pieces ; but the residence of

Mr. J. C. Baily suffered most, both as to itself and inmates,
Mr. Baily being found dead at the distance of one hundred yards, and his wife at the distance of three or four
hundred yards. This building encountered the fiercest
assault of the tornado, and was better calculated to test its
strength than either of the other mentioned, being new,
large, and built of heavy material, but was as a feather before it. Its lighter material, such as roofing, planking,
sash and window blinds, were carried far away, showers
of their fragments falling six miles distant, even of its
heaviest timbers, few were left near its former site, many
of them being thrown to a distance of hundreds of yards,
one in particular, a foundation sill, fifty feet long, and
some twelve inches square, passed above some buildings
two hundred yards distant and fell in a street in a broken,
shivered condition. A reasonable supposition is, that this
building, containing its doomed occupants, was lifted up
from its foundations entire, and torn to fragments as
hurried on by the whirling storm."

From the New York *Herald*, September, 1866.—" Tornado in New Jersey. A terrible and destructive tornado
occured near Mount Holly, on Friday evening last, the
14th inst. It commenced about a mile and a half from
Buddtown, uprooting trees, prostrating corn and fences,
doing comparatively no other damage until it reached the
residence of Hannah Alcott and sister, on the north road
from Mount Holly to Pemberton, a mile and a half from
the latter place. As it neared the residence of the Misses
Alcott, it prostrated everything before it, forest trees were
torn up or twisted off as though they were saplings,
fences laid low, and of 16 apple trees in the rear of the
premises, 14 were uprooted, and some of them moved a
distance of ten feet. The house, barn, wagon-house were
directly in its track, and they were all entirely destroyed
and thrown together, a perfect mass of debris. Such
utter destruction we never before witnessed. One of the
sisters was absent at the time, Hanna only being at home.
She had retired for the night, before the storm came up,
and when found was on her bed, which had been forced
against and among the branches of one of the large trees
prostrated near the fence. A heavy limb was directly
over her, serving as a protection from the broken debris,
furniture, &c. She was rescued from her perilous situa-

tion as soon as possible, probably having been lying there
for half an hour, and taken to a friend's house in Pemberton, when it was ascertained that her leg was broken in
two places, besides being dreadfully mangled, and her
head severely cut and bruised."

THE GULF HURRICANE OF OCTOBER, 1867.—From
the New York *Tribune.* The storm in the gu.'.. Along
the Rio Grande the hurricane was the most terrific within
the memory of man. Twenty-six persons were killed in
Matamoras, and 10 killed and one wounded at Brownsville. At Brazos, so far as known, 12 persons perished.
Two schooners were blown ashore. Only two houses
are left standing at Clarksville, and none at Bagdad.—
The loss of life at the latter place is not known. Ninety
of the inhabitants escaped by going on board vessels,
which rode out the storm ; the rest have perished. The
negro soldiers and their officers at Brazos are said to have
refused all assistance to struggling and suffering families,
and to have retired to a sheltered part of the island,
whence the soldiers returned next day to rob. One of
them was killed by a citizen who detected him in robbing.
Their officers, it is said, did nothing to check their bri-
gandism. At Brownsville the county court house and
jail was completely destroyed, and the prisoners are all at
large. The entire square, surrounded by Bown, Fort
Brown, 14th Levee and Elizabeth streets, were destroyed,
including the Ranchero office and Masonic Hall. The
Courier office had the roof blown off ; the Custom-house
wall was blown down ; the Presbyterian Church was destroyed ; the roof was blown off from the Post Office building. Between Tenth and First Streets seven brick and
four frame houses were rendered useless. These are only
specimens. In Matamoras 1,500 houses and huts were
blown down. The greatest distress prevails at all points.
Out of seven steamers only two can be repaired."

From the Whitby *Chronicle.*—A violent storm at Ux
bridge, which swept in a north westerly direction, for a
breadth of half a mile along' the Town Line, between
Uxbridge and Pickering. The storm was the most violent that ever visited the neighborhood. Considerable
damage was done to cattle and property. The houses and
barns of the following suffered more or less, some having
the roof totally blown off them : Messrs. Robert Spears,

Pews, Bar, and Rasnall, near the Town Line, and Mr.
Lamy, who had a mare and colt killed. Some idea can
be formed of the violence of the storm, while it lasted,
when it is stated that a harrow was blown half an acre
out of a field and landed on the road. Singularly enough
it is exactly twelve months ago since a similar storm, accompanied with hail, visited this locality, and completely
destroyed the growing crops at the time."

A FEARFUL STORM.—From the Weekly *Globe*, May,
1868. The telegraph despatches gave but a meager report of the remarkable and violent atmospheric commotion
which disturbed the inhabitants of Chicago on Tuesday
afternoon of last week. The city, suddenly, was at 5 p m.
plunged in utter darkness, and the citizens had to light
the gas in all the streets. The darkness was unusually
dense and chilly, giving a sensation as though a tremendous hail storm had passed very near. Three times did
the phenomenon appear and then pass away, leaving at
last the sun shining as brightly as ever. The telegraph
wires ceased to work at the time. In other parts of the
State this connection developed into an awful tornado.—
At Galesburg and Shangal a fearfi avastation happened,
with loss of life in the latter place, ...d the destruction of
fifty dwellings, school houses, and two churches. Service
commenced in the Advent Church, a new building completed last fall, at 4 p. m., as the people were wending
their way from their farms and cottages to church, the
sun was brightly shining, although clouds were seen in
the heavens. It was a day not now common in this
State, at this time of the year. The pastor of the Church,
G. W. Hurd, ascended the pulpit and commenced his dis-
course, which was not interrupted until it was nearly
through, when the evidence of the coming disaster began
to be apparent. First it was perfectly still, and then a
noise was heard in the distance as of the roaring of a
mighty cataract. The windows began to shake, and some
one called out from his seat, ' Mr. Hurd a bad storm is
coming up.' The minister answered, ' never mind the
storm, there is a day coming when there will be a storm
compared with which this will be nothing ; I will be
through soon.' Just then the hail and wind commenced
breaking in the window lights, and in almost an instant
the windows of the church, sash and all, were torn out.

The only two persons who succeeded in getting out were George Vern and Harrison Wixer, who were instantly killed. The building reeled like a drunken man, but none could get out. Wives clung to their husbands, children to their parents, brothers and sisters to each other, and despair was depicted upon every countenance.— Suddenly the crash came, and with a deafening sound mingled with the shrieks of the pent up people, timbers, scantling and all came down with a sudden crash upon the devoted heads of the congregation, men, women and children. Some had skulls broken, others arms, others received internal injuries from which they can never recover. Services were to have been held at the same time in the Methodist Church, but owing to the non-arrival of the minister the services were postponed. This church was also entirely demolished. So awe struck were the people of Chicago at the sudden apparition of darkness that in an editorial the *Republican* observes :—' Probably no one of the many persons enveloped by the darkness which fell upon this city with such mysterious swiftness on Tuesday afternoon but felt an indescribable awe at the sudden visitation. In one moment of time, without warning, as if the sun had suddenly expended its illuminating power, the light of day began to fade out of the sky, and night to descend upon the earth as if it were a mist,while a startling chillness permeated the air, as if the extinguishment of our central orb had instantly deprived our system of worlds of its boon of warmth. We may shudder when we contemplate what might have been the consequences. In the absence of any explanation from scientific sources, we presume the manifestation to have been those of a tornado whirling over Chicago."

THE GREAT STORM OF '67, ON THE ENGLISH COAST. —From the Edinburgh Daily *Review*. " The board of trade have received the annexed list of ships lost during the late succession of gales. The details are received or furnished by the receiver of wrecks, who adds the approximate value of each vessel and cargo. The ship Guy Mannering, 1160 tons burthen, Captain Brown, commander, from New York to Liverpool, totally lost on the rock at Iona (Scotland) seventeen of her crew were drowned, cargo very valuable : 1,600 bales of cotton, 5,360 barrels of flour, 38,986 bushels of corn and 40 cases of merchandise ; estimated loss of ship and cargo $200,000. The ship Severen, 856 tons burthen, owned by Messrs. Lidgett & Co., of Billiter St., London, from Calcutta for London, abandoned a total loss, her cargo was as follows :— 149,813 lbs of Assam teas, 226 tons of saltpeter, 750 bales of jute, 270 bales of hemp, 350 bales of safflower, 24 barrels of dye, 275 cwt. of turmeric, 13,572 hides, 400 tons of rapeseed, and 18 tons of linseed—crew saved ; estimated loss, ship and cargo, $340,000. The ship Simla, from Calcutta for London, burnt at sea, cargo: 300 tons of rice, 2,282 bales of jute, 150 bales of hemp, 7,800 hides, 36 tons of linseed, 350 tons of rapeseed, 58 barrels of dye wood ; estimated loss of ship and cargo, $250,000. The ship Albion, 1,245 tons burthen, owned by Messrs. W. Tapsott & Co., Liverpool, from New York to Liverpool, with general cargo, totally wrecked near Stromness harbor, eleven lives lost ; estimated value of ship and cargo, $200,000. The ship Attila, 945 tons burthen, owned by Messrs. John Treharne & Co., of Cardiff, totally lost off the coast of Ireland in coming from Quebec, estimated value of ship and cargo, $43,000. The ship Jalliet, 440 tons burthen, Captain Hitch, master, owned by Mr. Francis Chambers, of London, with a cargo of sugar and rum, bound from Demerara to London, lost at Hell Bay, Land's End, estimated loss of ship and cargo, $265,000. The barque Jeanne, 312 tons burthen, Captain Jaques Janssens, owned by Messrs. Chabot & Co., of Antwerp, bound from Buenos Ayres to Queenstown, wrecked near Maryport, estimated loss of ship and cargo, $170,000. The barque Lucetta, 192 tons register, from Seaham to Rochester, with coals, lost on the Gunfleet Sand ; value of ship and cargo, $18,500. The Diana, (barque) 261 tons register, of Liverpool, and owned by Mr. George Caldwell, of Troon, cargo coals, totally lost on the South point of Holy Island ; value of ship $20,000. The barque Norma, 650 tons burthen, Captain John Horstman, owned by Messrs. Konitsky & Co., Thiermann, of Bremen, with a cargo of tobacco, from Richmond for London, lost on the S. E. spit of the Goodwin; loss of ship and cargo $71,500. The barque Lexington, 344 tons, owned by Messrs. James Milligan, Jun., of Liverpool, cargo coals and machinery, bound from Liverpool to Havana ; totally lost one mile from Coral Point, N. W. of Islay, one man lost by remaining on board ; value of ship and cargo $40,000. The barque

Ariel, 365 tons, owned by Messrs. Robert Craggs and
Richards, and Richard Hall, of Stockton, with a cargo of
wheat from Odessa for Queenstown, abandoned thirty
miles from Old Head of Kinsale; estimated loss of ship
and cargo $47,500. The ship Eugenie, 1,136 tons, bur-
then, owned by Messrs. John Martin & Sons, of Dublin,
with a general cargo, from Liverpool to St. John's New
Brunswick, totally lost off Ballymac Cotter, County Cork,
captain and 12 men drowned; estimated loss of ship and
cargo $175,000. The barque Ayrshire, 681 tons register,
Captain William Henry Terry, master, owned by W. E.
Corner, of Leadenhall street, London, with a cargo of timber, bound from Quebec to Dundee, lost off Poor Island;
estimated loss of ship and cargo, $30,000. The Palinurus,
1,082 tons, Captain Andrew Berry, owners Messrs. R.
Allen & Co., of Liverpool, with a cargo of coals from that
port for New York, wrecked at Cymyran Bay, one life
lost; estimated loss of ship and cargo, $78,000. The above
is scarcely a tithe of the ships that have been lost; they
number between three and four hundred. The gales
were very severe, and see what loss in this small item
alone, it is $1.948,500.

Now my kind reader, in overhauling those items, I do
it in love to truth, its Author, and my fellow traveller to
the judgment. God, in giving you and me tokens, as well
as to the world, wants us all to take heed, for he does
nothing in vain. What then does he say, see Jeremiah
xxx, 23, 24; hear it: " Behold, the whirlwind of the
Lord goeth forth with fury, a continuing whirlwind, it
shall fall with pain upon the head of the wicked. The
fierce anger of the Lord shall not return, until he have
done it, and until he have performed the interests of his
heart; in the latter days ye shall consider it." Does God
mean what he says? if so, it is our exalted privilege, to
take heed. And whether or not reader, you can see by
causes and their effect that we are nearing the end of this
age I am satisfied; but it shall be as it was in the days of
Noah and Lot, none of the wicked understood, and the
reason was, they would neither believe Noah or his preaching, nor Lot or his solicitations. And Christ said in the
xxiv of Matthew, that as it was in the days of Noah and
Lot, even thus shall it be when he comes, and I believe
it. We shall give a little further history of the hurricanes
before we come to the chapter on earthquakes.

Weekly *Globe*. July, 1866.—"The Hurricane in Halton.
—A correspondent informs us that on Friday evening 13th
inst., part of the County of Halton was visited with a
hurricane and hailstorm, that had no parallel in the
memory of the oldest inhabitant. The course of the tem-
pest was about due South, perhaps varying to the southeast, and it carried a width of two miles, though its main
fury was poured out in the centre mile. It seems to have
mustered all its forces in the immediate vicinity of the
Hornbys. East and West, to the North of that its power
was but little felt, except in the woods of R. Howson and
J. Taylor; but to the South of Hornby, for four miles, it
has spread wreck and ruin all around, and I have heard
that it extends to the lake. It began raining about half-
past eight o'clock in the evening, having, to my eye, the
look of an ordinary thunder storm, with a great deal of
lightning, which continued for about 20 minutes, when
the rain poured down in torrents, accompanied by a fearful hail, and the wind rose to a pitch we have never
known before. The hurricane lasted only twenty or
thirty minutes. No harm came of the lightning, nor was
the thunder at all loud, I have heard it twice as loud.
The hail did considerable damage to the crops and windows. It was of a diamond shape, three quarters of an
inch by one-third, or, say the size of a pigeon's egg.
Robert Hall Esq., had 45 panes of glass broken; half his
apples were knocked off, and so abundant was the hail,
that a large quantity could have been shovelled up by the
side of his buildings. About a mile from that, at Hornby,
I saw oats from the fields of Mr. Pickard quite destroyed.
The hurricane, in many places, partook of the nature of
a whirlwind, and its full force was exerted at 15 or 20
feet from the earth and upwards. Its fury was terrific,
barns and sheds were crushed before it as you would
break an empty egg-shell in your hand, their roofs scattered to the winds. I saw rafters carried 300 yards. You
could see its power in the forest and among the fences on
every hand; one-half the timber is in places thrown down.
The fine brick Wesleyan church at Hornby is in part unroofed and so scattered as to be unfit for repair. A con-
cession to the south-west of Hornby, and for a length of
two miles, has suffered most. Mr. T. Chisholm had a new
frame barn just finished, completely demolished; opposite,
Mr. Noakes had a barn and dwelling house unroofed.

On the next two farms, Mr. Cunningham had all his out-buildings unroofed, and Mr. Irvine a shed levelled, the next farm escaped, the buildings being under range, but on the next, Mr. Campbell had a barn and shed unroofed. Mr. Hempstreet, on the next farm suffered most, fences and trees scattered broad-cast; about half his fine orchard was uprooted, and his barn was at one end, carried bodily 40 feet, at the other, 36; a large addition to his barn was levelled, and all his buildings destroyed, save one protected by the rest. Jonathan Hows, on the 7th line in Trafalgar, had a new gravel house just erected, completely crushed. Not a single house in which any one lived has received more than a trifling injury. I heard of one narrow escape. The wife and children of J. Anderson, on the 5th line, were in their sitting-room, and they had no sooner left it than the roof of a shed came crash through the sitting-room, breaking the bureau, etc."

THE CONVULSION OF THE SANDWICH ISLANDS —From the New York *Herald*, April, 1858:—" The greatest volcanic eruption recorded in modern times has occurred on the Island of Hawaii, one of the group of Sandwich Islands. For some time past it has been observed that the crater of Kilauea was very active, and that a new volcano had been formed. The volcano is the well known Manna Loa, and it has an elevation of 13,758 feet. On the 27th of March last the new eruption commenced and has continued up to the latest dates. During twelve days there have been 2,000 shocks of earthquake, followed by fearful tidal waves which have destroyed entire villages, and caused the death of 100 persons. For fourteen days the district Kona has been the centre of motion for the great eruption. A gigantic stream of molten lava is flowing from the summit of Manna Loa across the lands of Kakuka and Poakini to the sea at Kaaluala landing. The slope and part of the summit of a mountain fifteen hundred feet high have been lifted up bodily by the earthquake and thrown over the tops of trees for a distance of over 1,000 feet. At Wahoinee a creek has opened, extending from the sea. To as high as the eye can reach on the slope of Manna Loa the lava is from one to seven feet in width, and an eruption of moist clay was thrown from the side of the mountain, between Lyman's and Richardson's, a distance of two miles and three quarters, with a width of one mile, in the short space of three minutes. This

7

terible eruption overwhelmed houses, persons, and hundreds of animals, and scattered death and destruction wherever the clay fell. A column of smoke seven and four-fifths miles in altitude was thrown out of Mauna Loa obscuring every thing for miles around, save where the bright spiral pillars of fire flashed upwards from the mouth of the volcano. The sight was one of the grandest but most appaling ever witnessed and almost defies description. The immense tidal waves rushing in with so great a height that they swept over the tops of the cocoanut trees on the Kona coast. During the severe shock of earthquake, which took place on the 2nd of April, no living creature could stand for a moment. Immense bodies of earth were tossed about at great distance, as if they were feathers waved from point to point by a storm of wind. No one stone stands upon another as before in this district. Immense precipices which have hitherto been a terror to all who have seen them, have been leveled to the earth, and where the ground was formerly smooth and unbroken for miles around, the earth has been rent asunder and upheaved, forming gigantic chasms and precipices. The entire topographical appearance of the country has been so completely changed that even those who have lived in the desolated district all their lives, cannot recognize it, or point out localities with which they were formerly familiar. Luckily this part of the island is but sparsely populated, and the lands are not in general cultivation. The loss of life as far as can be ascertained as follows: In the village of Pilinka, 33; at Mokaka, 13; at Palalua, 4; at Hona, 27; at Vanilo, 3. This makes a total of 80 persons killed as reported up to the present time. There are rumors that the casualities considerably exceed 100. All of the unfortunate persons who have lost their lives, were native Hawains, not a white person being killed, or in any way injured. At the present moment the entire group of islands is enveloped in a dense black smoke, and the indications are that Mauna Loa is still in active volcanic eruption. A vessel has just arrived from Hawaii, bringing later accounts of the lava flow, and of the eruption in general. The first stream of lava broke out from the crater of Mauna Loa, some 2 miles above the residence of Captain Robert Brown, and flowed directly towards it. It came down the mountain side in a broad stream, several feet in depth, and travelled with such a rapidity that the family in the house had barely time to escape, taking away with them nothing but their clothes; the path they took was perfectly free from lava, but ten minutes after they left it, and reached a point of safety, the entire road was covered with the fiery stream. The lava pushed onwards to the sea, and drove the water back with such a violence that it became agitated and convulsed, and huge waves rolled towards the ocean as if lashed to fury by a storm. The ground thus occupied is now a mass of lava, forming a point for at least one mile in length, and as the stream continues to descend the probability is that it will remain stationery, and form a portion of the island. The most terrible shock of earthquake, which took place on April 2nd, burst open the earth at the village of Walschina, and a tidal wave rushed inwards with fearful effect; it was over fifty feet in height, and swept over the tops of the cocoanut trees, carrying death and destruction to persons and property. Throughout the island this shock was felt with fearful effect; buildings of all kinds were torn from their foundations and hurled great distances, and many persons and animals lost their lives. The scene at the craters was appaling; huge rocks were hurled from their mouths, accompanied by streams of lava hot and red, which attained an altitude of 1.000 feet. When it fell it rushed down the mountain towards the sea at the rate of ten miles per hour. The new crater which was formed on the 27th of March, is over two miles in circumference. It vomited rocks and broad streams of liquid fire which illuminated the night for an area of over fifty miles. In addition to the one mile of land formed by the lava, driving back the sea, another stream extending for a distance of three miles poured down the mountain striking the water with a tremendous shock. At this time another earthquake shock occured, and immediately after an island four hundred feet in height, rose above the water, and was soon after joined to the island of Hawaii by the stream of lava. The eruption of moist red clay took place during the great earthquake shock, and went rushing across the plain below for a distance of three miles. From the midst of the crater, from whence this came, an immense stream of water is now pouring down. The entire section of country around Manna Loa has been desolated. A stream of lava is flowing under the ground, six miles from the sea, and has broken out in four places, each throwing up brilliant jets of fire. The base of the volcano is about thirty miles in circumference and now presents a most barren and desolate aspect, the gases rising from the rent earth having completely destroyed all vegetation. The earthquake shocks were felt in all the Sandwich Islands, but only around Manna Loa was the effect disastrous.— The eruption still continues with unabated violence, and the scene is one of the most terribly grand that has ever been witnessed. From the crater of Manna Loa the huge column of smoke continues to ascend, hiding from view the skies and clouds, and enveloping the entire country in partial darkness. Every now and then thick streams of lava shoot upwards from the midst of the smoke, illuminating everything around. Hundreds of jets of flames burst from the lava and are thrown for a distance of a thousand feet, the whole forming a pyrotechnic display of surpassing magnificence. Shock after shock of earthquake convulses the island, and ever and anon the low rumbling sound which breaks out from amid the din and noise of the eruption, indicates where the earth has been violently torn asunder, or where the summits of huge hills and mountains have been hurled from their places and sent rolling downwards to their base." Reader I must bring this chapter to a close, although I could write volumes on this mighty sign alone. The Prophet Isaiah, in chapter xxiv, 1, 7: "Behold the Lord maketh the earth empty, and maketh it waste, and turnith it upside down, and scattereth abroad the inhabitants thereof. The earth also is defiled under the inhabitants thereof, because they" (the inhabitants) "have transgressed the laws," (the second and fourth commandments) " changed the ordinance," (baptism) " broken the everlasting covenant," (the Sabbath.) Blessed are they that do his commandments. See Rev. xxii, 14.

How sweet to reflect on those joys that await me
In yon blessed regions, the heav'n of rest,
Where glorified beings with welcome shall greet me,
And lead me to Mansions prepar'd for the bless'd.

Encircled with light and with glory enshrouded,
My happiness perfect, my mind sky unclouded;
All rage in the ocean of pleasures unbounded,
And range with delight through the Eden of love.

Though prison'd on earth, yet by anticipation,
Already my soul feels a sweet prelibation
Of joys that awaits me when freed from probation,
My heart now's in heaven, the Eden of love.

Then songs to the Lamb shall re-echo through heaven,
My soul shall respond to Emmanuel be given
All glory, all honor, all might and dominion,
Who brought us through love to the Eden of bliss.

1. Don't you see my Jesus coming,
    Don't you see him in yonder cloud,
    With ten thousand angels round him,
    See how they my Jesus crowd.

2. Don't you see the saints ascending,
    Hear them shouting through the air ;
    Jesus smiling, trumpet sounding,
    Now his glory they shall share,

3. Dont you see the heavens open'd,
    And the saints in glory there ;
    Shouts of triumph bursting round you,
    Glory, glory, glory here.

4. Come back-slider though you have pierced him,
    And have caus'd his church to mourn,
    You may yet regain free pardon,
    If you will to him return.

5. Now behold each loving spirit,
    Shout the praise of his dear name ;
    View the smiles of their dear Jesus,
    While his presence feeds the flame.

6. There we'll range the fields of pleasure,
    By our dear Redemer's side,
    Shouting glory, glory, glory,
    While eternal ages glide.

The reader, I hope, will pardon my little bits of poetry.
We shall now give a brief account of earthquakes.

CHAPTER VII.

EARTHQUAKES.

From the Three Rivers *Inquirer*, January, 1866.—"A
curious convulsion of nature occured lately at Bon Desir,
county of Tadonoac, C. E., by which an immense marl
hill, a dwelling house, barn, and a number of other

buildings were removed to a distance of two acres below their original sites, and the beach close by strewn with immense boulders and raised to a height of 30 feet above its ordinary level, while the ground all around over an area of 13 acres, was cut up with deep crevices. The slide covered over 12 acres in width by five in depth."

From the Weekly *Globe*, January, 1866.—"An earthquake occured on the 3rd inst., in Mexico, doing considerable damage at Orizana, Maltida, and other places in the interior with loss of life. In a recent issue we stated that a shock of earthquake had been felt on Friday morning, the 15th inst., at Lake Beauport, we since learn that, at precisely the same hour, the shock was felt at Grande Baie and Ha Ha Bay, on the Saguenay. It appears to have been a pretty lively concussion and to have affected nearly the whole of the north shore, having been also felt at Baie St. Paul. These shocks have not been unfrequent lately in the Lamentide range."—*Quebec Chronicle*.

From the London *Times*.—"Earthquake. Tenacity of Life. At the Royal Institution, London, Dr. Laeaita recently delivered a lecture on the earthquakes of Southern Italy, and stated that during the last seventy-five years the Kingdom of Naples had lost 110,000 inhabitants by such calamities. In 1783 a young and beautiful girl was buried under some ruins caused by a great earthquake, and was dug out alive after eleven days, during which she had counted the days by a single ray of light which reached her through a crevice. She lived for nine years after, but was always sad and gloomy."

From the London *Times*, 1867.—Terrific earthquake in Egypt. At Alexandria, and throughout Egypt, earthquakes have been felt. At Cairo 200 houses have been overthrown, 300,000 inhabitants have taken to the fields and encamped. Smyrna and the Greek Archipelago were shaken by simultaneous shocks. The Isle of Rodes has been devasted, and one of its largest forts rent open, and thrown down. There was a violent shock of an earthquake on the 12th at Malta, which has seriously damaged the fortifications. At Corea, the capital of the Isle of Candia, the destruction has been enormous. The city and mole are partially destroyed. An earthquake has occured at Naples. It was horizontal and came along the coast, and was so violent that it rang bells, opened doors, and shook the beds so violently tnat the soundest sleepers were awakened; many people spent the night in the streets, and amongst others, the Count of Syracuse, who sat in his carriage. At Sorento the people were all equally alarmed, and many of them spent the night in the streets. The Madona was carried in procession all around the Magellina, and propitiation thus offered to heaven."

From the *Signal*, 1868—The earthquake in South America. Although over a montl. has elapsed since the first and fatal shock, yet in Arica, Arequipa, and many of the interior ruined towns, the ground still cracks and jars, and trembles, and the shocks have been counted by hundreds already with no settled stillness so far. The poor affrighted people are escaping by every steamer coming to Callao, and those remaining in the mountains or scattered over the pampas, are afraid to re-build the fallen ruins, owing to the continued heaving and uprising of the ground. How much longer this will continue no one can determine, but many think it is the indication of another fearful outbreak."

From the Weekly *Globe* of Nov., 1868.—" We have had of late a large number of earthquakes in different parts of the world, varied by tempests of unusual severity, and floods by which multitudes have perished. In the West Indies and South America, as we all know, these earthquakes have been peculiarly disastrous, and since the time of their occurence news has reached us of something of the same sort in different quarters in both hemispheres. The last accounts speak of destructive winds and storms which have swept over Northwestern Mexico. One town in the State of Sonora is said to have been entirely destroyed by floods and whirlwinds. Whole herds of cattle have been, it is said, swept away, and different crops totally ruined. It naturally rises, what may have been the reason for all this unwonted elemental activity ?" A question, says the *Globe*, much more easily asked than answered. Reader, I shall allow Gods word to answer the above question. Read Jeremiah xxx, 23, 24 ; Matthew xxiv, 7. I shall make no comment on these texts, only ask my intelligent reader to refer to them.

From the Weekly *Globe*, Nov., 1868.—Naples, Nov. 20th. The eruption of Mount Vesuvius is still very threatening ; all the cones of the volcano emit rumbling sounds, and eject large quantities of lava. The lava has set a whole forest of chestnut trees on fire, causing immense devastation. Houses, farms and lands are overwhelmed with ruin, and the population of the nearer villages are leaving their houses in great destitution."

Naples, Nov., 21.—" The eruption of Mount Vesuvius is increasing in violence. Many houses and farms in the vicinity have been utterly destroyed. The market town of San Giorgo is in imminent danger. The ship Imperial from Kodiac, reports a violent shock of an earthquake at that place at 8 p. m., of September 5th. Three houses and nearly all the chimneys in town were shaken down. The sensation on the ship was terrific. She seemed as though passing over the rocks at great speed, while articles were shaken down which the most violent gale had not disturbed. In the southern parts of the island large rocks were torn up and thrown down the mountain. The shock lasted forty seconds."

The Paris correspondent of the London Times, 1867.— " At about a quarter past five yesterday morning a shock of an earthquake was felt in Paris and its neighborhood, especially in the direction of Versailles, in which two persons say they were awakened by a motion, they heard a cracking of the walls and floors, and that the first shock was followed by several others. Persons in Paris have told me they awoke about the time the shock took place, but went asleep again without being aware of what had occured. It appears that it was also felt in the departments chiefly, so far as is yet known, in the west, centre and south west of France. The shocks are estimated to have taken eight or ten seconds. At Limoges it seems to have been severe and accompanied by a noise compared to that of trains passing through a tunnel, and in the houses, according to letters received, the beds moved, the crockery and glass clattered, the bells rang, and the inhabitants were all on foot. In the neighborhood of Paris I know of persons who got out of bed in alarm, thoroughly roused by the first shock. At Dourdon the church bells rang spontaneously at five in the morning. There the cure, not suspecting an earthquake, unhesitatingly attributed t! ̄ phenomenon to evil spirits, and proceeded forthwith to the belfry with bell, book and candle to exorcise them. The oscillatory motion of the earthquake, as

already mentioned, was from northeast to southwest. It
followed very closely the Western Central and South-
western Railway lines. News of it comes from Blois,
Saumur, Angers, and Nantes, from Bourges and Limoges,
and from Poitiers, Mart and Angouleme. There is no
instance on record of a previous earthquake in any of these
districts. A great many people thought the end of the
world was come." It surely will come, and that soon, but
none of the wicked shall understand. "It is stated that
two distinct shocks were felt in Devonshire, England, a
few hours before that in Paris."

From the Weekly *Globe*, Oct., 1868.—San Francisco,
Oct. 21.—"A heavy shock of earthquake was felt here at
7:50 a. m. Several buildings on Pine, Battery, and San-
son streets were thrown down and a considerable number badly damaged. The ground settled, which threw
buildings out of line, and at present, 9 a.m. no estimate
can be made. Several shocks have followed at intervals
since, creating a general alarm. A shock was felt with
great severity at San Jose, where a number of buildings
are considerably injured. A survey of the city shows that
the principal damage by the earthquake is confined to the
lower portion below Montgomery and among the old
buildings on the Made Ground. Numerous houses in
that portion of the city have been abandoned and pulled
down. The Custom-house is considered unsafe, the offi-
cials have removed from it; business in the lower part of
the city is suspended, the parapets have been thrown
down, walls and chimneys, causing loss of life. At Oak-
land the shock was very severe; the ground opened in
several places and a strong sulphurous smell was noticed
after the shock. The Court House at San Leandro, was
demolished and one life lost. From various portions of
the country and in the vicinity of San Francisco Bay, the
shocks are reported as severe, and in many places the
earth opened and water gushed forth. Twelve shocks
were felt during the day. The greatest damage exists on
a hill several hundred feet wide, running about north-
west and southeast, commencing near the Custom-house
and ending at Falsom street wharf, injuring and demolishing about twelve buildings in its course. At the corner
of Market and First streets the ground opened several
inches wide and about fifty feet long; in other places the

ground opened and water was forced above the surface.—
The City Hall is a perfect wreck. The courts have all
adjourned. The United States Marine Hospital, United
States mint, Lincoln school house, San Francisco gas-
works and deaf and dumb asylum are all damaged. The
shock was felt about the shipping in the harbor as if vessels had struck upon a rock, although the water was perfectly smooth at the time of the shock. Shocks were felt
at Sacramento and Stockton. The Central Coast and
Almeda Companies' building was thrown down and some
lives lost. At Red Wood City the large brick court house
is demolished. Another shock has just been felt. Private
despatches from San Francisco say the loss thus far will
exceed $300,000. A number of buildings have sunk
several inches, and streets before level are uneven in
many places. A ship anchored fifteen miles outside felt
the shock very heavily, and it was the same with vessels
on the bay. Many casualties occured from frantic efforts
to get out of the buildings ; some persons jumped out of
the windows in the second story. This vicinity appears
to have been about the centre of the convulsion. South
and east, persons in the country noticed an upheaving
and opening of the earth in several places. Large masses
of rocks, several tons in weight, were detached and rolled
down the hill, and mountain side. Persons here who
were in Peru during the late earthquake, state the shock
yesterday to have been as strong as some of those that destroyed so much life and property there, and say that a
recurrence of the shock was all that was required to make
the disaster as great. $3,000,000 will not cover the loss
by damage to the buildings alone. There are six killed
and many wounded."

San Francisco, Oct. 23 —From the exterior we learn
Almeda county suffered most by the earthquake. Fissures
in the earth were made, from which issued clouds of dust
and volumes of water. Creeks dry for several months
suddenly became large streams ; hot water and steam also
gushed from the earth. The villages of San Leandro and
Hayneard are almost in ruins. The brick buildings were
all thrown down, and hundreds of tenements rendered
unhabitable. The towns of Almeda, Brooklyn, and Oak-
land suffered severely, as did San Jose and Redwood city.
The brick buildings in the old mission of San Jose are a

mass of ruins. The damage in Petaluma, Healdsburg, Santa Rosa, Valjo, and Martinez, was considerable."

From the London *Times*. 1867.—"From statistics obtained by reliable persons, it seems to be proved that these upheavals of the crust of the earth, whatever their origin, are greatly increasing in number and violence. Among the earliest quakes recorded is that by which Herculaneum and Pompeii were destroyed in the year '63. In 526, Antioch, in Syria, was almost entirely destroyed, a number of persons perishing in the ruins, being estimated at one quarter of a million. The most memorable earthquakes in history are as follows : In 1692, Port Royal, the capital of Jamaica, was entirely submerged by the force of an earthquake, which swallowed up over a thousand acres, and drove ships so far inland that they floated above the buried city. In 1772, an entire volcano sank into the earth in the Island of Java, carrying with it forty villages; the mountain itself, which was fifteen miles long and six broad, accompanying the hamlets and their 2,957 inhabitants. On the 1st of November, 1755, occurred the memorable earthquake of Lisbon, by which 60,000 perished in the twinkling of an eye. Here also was the great tidal wave seen of an altitude of fifty feet. One of the most awful incidents of this earthquake was the sinking of the city quay. This had just been constructed of marble at an immense expense, and to it as to a last refuge, fled thousands of the hapless inhabitants. Without a moment's warning, the earth suddenly opened to receive it, and after sucking in the mass must have closed over it, as not a single body of all the thousands that went down, nor the least spar or ark from any of the ships near by that were sucked into the chasm ever came to top. The water there is nearly 600 fathoms deep, and at an unknown distance beneath the bottom repose the helpless Lisbonese. This Lisbon Earthquake Humboldt estimates effected a portion of the earth four times as large as Europe, and was felt in the Alps, on the coast of Sweden, in the West Indies, on Lake Ontario, and along the coast of the Massachusetts. In 1811 the earthquake on the Mississippi, severest at New Madrid, Mo., shook the ground for many days, and alternately raised and depressed it here and there, the latter sections forming a section called the Sunken country to this day. On the 26th of March, 1842, a violent thunderstorm, with incessant flashes, was observed by the people of New Madrid, and at the same time the city of Caracus, in South America, was laid in ruins, 12,000 of its people perishing. The great eruption in Vesuvius in 1857, with accompanying earthquakes, will also be remembered as leading to an immense destruction of human life, variously estimated at from 22,000 to 40,000 souls. In 1858, June 19, the valley of Mexico was also devasted by one of these visitations, demolishing houses throughout its length, and destroying the costly aqueduct supplying the city with water. March 22, 1859, Quito, in Ecuador, was nearly destroyed by an earthquake and thousands of lives lost."

From the Weekly *Globe*, September 18th, 1868.—" The most terrible earthquake that ever occurred on the coast of South America took place on the 13th ult. At first but little was known of the dreadful catastrophe that had visited with deadly effect nearly every city on the coast. The events of the 18th and 14th ult., at Callao, of which you have been already informed, was but the premonitery revelations in the great disaster that had occurred. The sea it was known, had been greatly agitated, and risen to a great height, and inundated part of the city, but, beyond driving many persons from their homes in the terror which the fury of the elements is well calculated to inspire, it was believed no serious disaster had occurred ; but it was found that the whole western coast of South America had been visited by the most terrible earthquake that has ever occurred, and that eight cities, among the most important on the coast had ceased to exist. Many more cities are reported to have been more or less effected, and its effects were experienced at many points northward from Arica to Callao, a distance of 650 miles, and southward to Cobija, a distance of 280 miles. Arica is the central point from which the effects of this disaster must be traced. It is a seaport town of Peru, with a population of about 30,-000 ; but, though a better landing place than most of the contiguous ports, owing to the heavy surf it has always been difficult and hazardous for shipping. About 5 o'clock in the afternoon, in the most mountainous part of the country back of Arica, a sensation was experienced like the collision of two heavy masses, and from this point radiated North and South to the distance I have already indicated, with consequences as terrible as they were immediate Three undulations of the earth followed this unseen convulsion of nature, and each undulation was accompanied by a tidal wave, the second of greater lateral extent than the first, and the third greater than the second. Before proceeding to give any account of the disaster in other places, it may be well to relate all that I have seen, or have been able to gather of the character and extent of the catastrophe at Arica. A gentleman who was at that place when the disaster occurred gives a vivid description of the scene. He says, the hour was that when by custom most of the inhabitants had just closed their labors and at their homes, the instant the startling indications of an earthquake were felt, there was a general rush for uncovered spaces, which were reached by many uninjured, but not by all. The streets became a scene of terror; all the houses in the city trembled like a leaf, then they surged, and some of them fell to pieces with crash after crash. At this juncture, when the undulations were active, the earth opened in several places in long and almost regular lines. The fissures were from one to three inches in width. The sensation was distinct as though something was rolling underneath. From every fissure there belched forth dry earth like dust, which was followed by stifling gas. Owing to the demolition of buildings and the general destruction of all kinds of property, and the dust belched forth as well as that set in motion by the general tumult, a dense cloud was formed over the city that obscured the light. Beneath the cloud was the gas, which severely oppressed every living creature, and would have suffocated all these if it had lingered longer stationary than it did, which was only about 90 seconds. The undulations were three in number. Each succeeding one was of greater magnitude than the former. When the undulations ceased, the cloud of dust ascended and dispersed, and light was restored, then quakes at short intervals succeeded, as though subterranean explosions or collisions were taking place; at this time people from all parts of the city fled to hills, amid falling stones and timbers, which descended from swaying walls and broadly rent buildings on the eve of crumbling into perfect ruin. Some were struck dead by the falling materials, and others. were maimed, while all were made to staggar from side to side like people in a state of intoxication. Many of both

sexes carried children in their arms, and those who had
met these, articles of value ; the avarice of some was stronger
than fear, even amid this terrible convulsion, and hence
there were those who dallied to collect valuables, many of
them who suffered for temerity, either by the sacrifice of
their lives, or otherwise. As the rush for the hills continued, and stones and materials, of all kinds were falling
and the houses crushing, the people were struck down,
and either killed or dangerously hurt The water in the
harbour was now receding from the shore, bearing with it
all the shipping at rapid speed, then the current changed,
and before an almost overhanging, tremendous wave, the
vessels came back, tossed one way and then another, or
whirled about as though they were only floating logs, and
on the very summit of this immense volume of water
rode the United States steamer Wateree. The huge wave
dashed against the stone moles or pier, and shattered it all
to pieces, then swept from its path what was standing of
the custom house, and almost every vestige of the ruins of
other buildings ; it rolled over the already destroyed
houses of the city, and set a myriad of articles afloat which
eddied in every direction, while at the same time the
vessels and floating materials were forced ahead of the
waves at this time curling and foaming summit, every
thing which it encountered in its course was swept away
in an instant ; even great masses of stones were rolled over
and over. When the force of the waves were spent, it retired, and in a short time the equilibrium of the water
was restored, and then it occupied about the same time
and presented nearly the same appearance as it did before
the earthquake. The vessels carried inland were a terrific sight. The most of them were bottom upward.
Their masts had been snapped like sticks. All this dreadful picture the refugees on the hills were now beholding.
Many of them did so with the most marked evidences of
fear, because quakes of the earth were still felt at short
intervals. The Wateree was grounded inland at a quarter
of a mile from the beach, on the line of a railroad. Near
the Wateree a Peruvian war vessel was also grounded and
so gently that none of its rigging nor any of its timbers
were impaired. On board the Peruvian war steamer 80
lives were reported to have been lost. The United States
storeship Fredonia was upset, and all her crew were
drowned according to report. Those who escaped were

the captain, surgeon, and a paymaster, who were on shore
when the earthquake occurred, and sought refuge on the
hills. A British vessel called the Chanarellic lost many of
her crew, who were tumbled overboard. A United States
brig, name not known, was foundered with all on board.
The vessel, it is reported, was laden with guano. The
rest of the shipping destroyed were South America coas-
ters. The fatal casualties in the city were about 50, and
the other casualties about 100. The total loss on ship-
board was about 500, principally fatal. The refugees remained on the hills for two days, during which time they
suffered greatly for food; at the same time the quakes continued as before, at intervals. In time their suffering became so intense that the males had to go in search of food,
a limited quantity of which they found in a damaged condition, and with this they succored their families. The
second morning after the earthquake, a light draught
coasting vessel entered the harbor, but did not remain
long at anchor. As soon as the captain became aware of
what had happened, he took on board a number of people,
and left with them for Callao; thence one of the parties
got to Panama. It was impossible for any one to stand,
men fell as though they were intoxicated. The shock was
so severe as to cause the earth to roll from side to side,
so that the bells of the churches were set in motion
and chimed forth doleful peals: the houses rocked from
side to side, the earth rose and fell, and all the motions of
a steamer in rough weather were experienced. After
this frequent shocks were felt, and the sea began to leave
the land about the same time and in the same manner as
at other points, until 10 o'clock, when a complete inundation took place. The port of Cerro-Azal was destroyed
by the inundation. The loss is over $50,000. The beau-
tiful city of Arequipa is completely destroyed, not a church
has been left standing nor a house habitable. The houses
being very solidly built and only one story high, resisted
for a few minutes, affording time for many of the inhabitants to get into the streets, so that the mortality, although
great, is not so considerable as it might have been. But
at least 2,600 persons perished. The convicts in the public prisons and the sick in the hospitals were all crushed
to death. The shrieks of the men, women and children,
frantic with fright, the crash of falling masonry the upheaving of the earth, and the clouds of burning and sul-

locating dust, altogether constituted a scene which baffles description. The earth continued in motion for eighteen hours, and slight shocks are still felt. Chala is a heap of ruins. The inhabitants saved themselves by running to the mountain. The town of Tambo is also washed away, and it is reported that 500 persons perished. The towns of Trobaja, Vitar, Mollendo, and Megla, and over 150 miles around were completely destroyed. In the two latter places the material buildings of the Arequipa Railroad were deposited on the ground, all of which were swept away. In fact the whole coast south of Callao, as far as Iquique, is one mass of ruins. The number of lives lost cannot yet be rightly ascertained, but must be very heavy. The loss to Peru by this terrible visitation is beyond all calculation, and the misery and desolation which must follow in the wake of this catastrophe is beyond all description."

CHINCHA ISLANDS.—On the morning of the 13th, the sun shone brighter than it was ever known to do in this season of the year, calling forth remarks from all persons as a strange and unusual sight. A strong but pleasant breeze was blowing at the time, arguing fine weather.— Nevertheless, on the north part of the Island, at a long distance off, a reddish vapor was seen arising out of the sea, and at 12:20 p. m., a strong wind sprung up from the south almost approaching a hurricane, which lasted until 4:38 p. m., after which a prolonged and distant noise was heard resembling thunder. A complete calm ensued immediately, after which the first shock of earthquake took place. This lasted for four minutes 18 seconds; so great was the motion that people were thrown down. After this a mighty crash was heard, as though a powerful wave had broken on the rocks; but the sea was calm. At 5:56 p. m., the earth again began to tremble, and continued to do so for two and a half hours. The sea now became entirely calm, but the birds were seen abandoning the sea and rocks and soaring to the element above, screeching most horribly, as though they were aware of what was about to take place. The night became pitch dark, nothing could be seen over sea or land, the breeze felt during the day began to blow, adding further terror to the people who were momently expecting the island to be swallowed up. At 9:45 p. m., persons living in the vicinity of the mole noticed that the water was leaving the land, and the alarm was immediately given, *se sale el mar*. Nothing can describe the terror which prevailed on the north island.— The population, consisting of from 500 to 600 souls, all forsook their houses, and took refuge on the highest part of the island, some naked, others sick and infirm, women and children, all thronged in one spot, dumb with terror. At 10 p.m., the breeze lulled and the heavens cleared so that a view could be obtained of the sea, which had retired from the land about 70 yards. This circumstance is the more remarkable from the fact that the depth of the water in the bay at low tide is from 15 to 20 fathoms, but soon after the water was seen to rise iO the shape of a colossal wave, and in less time than it takes to describe it it had almost covered the island, washing away the houses and everything it came in contact with. The losses are heavy. Both moles are destroyed, many launches broken to pieces, and the houses in the vicinity washed away. In one of these was a family of three, husband, wife and child. My informant says that after the sea became quiet he procured a boat and proceeded to the other two islands to ascertain the damage done to the shipping. He said : '.When we arrived at the foot of the channel, we heard a noise heartrending and doleful. The sea rose and fell at the rate of ten feet in every three or four minutes, and I saw the vessels in awful confusion. Some had parted their chains and were drifting on the rocks, and others endeavored to get under way. I could not force my men to proceed further, and we were compelled to return. I am unable to give your readers an entire and correct list of the vessels injured and lost, but the following are among those which have suffered the most :—English ship Resolute, partly destroyed; English ship Eastern Empire, greatly damaged ; English ship Royal Oak, in a very bad state ; Prussian barque Leopold Hod, an entire wreck ; English ship Oceanica, nearly an entire wreck ; English ship Southern Ocean, greatly damaged, and also, American ship Shatemuc. All the ships have suffered more or less, as will compel them to go into repairs."

GUAGAQUI, August 26.—We are in the midst of horrors. The mail from the capital arrived here yesterday afternoon, and has brought frightful news. A terrific earthquake took place at Quito, at 1:20 a. m., on the 16th inst.,

which extended in a greater or less degree over the whole
of the northern part of the Republic. The loss of life is
frightful; it is estimated that nearly 80,000 persons perished. The towns of Ibarra, Otavalo, San Antoni, and Rini-
co, and numberless farms, are wiped off the face of the
earth. The suffering of these poor people, who are wand-
ering over the country, without money, without food and
clothing, cannot be described. Business of all kind is totally suspended. The panic is so great that the people
think only to save themselves from impending destruction,
hence a terrible picture will soon be presented—actual
famine and starvation must ensue. The details thus far
to hand are harrowing enough, but worse must come.—
The following is a translation of a communication from the
Governor of the Province of Imbabura, received here
yesterday. It is dated

IBARRA, August 17, 1868.—" In the midst of the most
profound consternation, which has filled the few who
have escaped the complete destruction and ruin of this
town, I have to inform you that on Sunday, the 16th
inst., at 1 o'clock in the morning, the entire town of Ibarra was buried in its own ruins. Induced by a terrible
earthquake, originating, it is believed, in the volcano
Ocampo of this town. There remains nothing but shape-
less ruins, and but one sixth of the population survive.—
Of those who remain alive the greater portion are maimed and injured for life. All the towns in this vicinity
have likewise been destroyed. The shocks continue to
this writing, being repeated every hour." Reader, the
sad picture presented to our view in this calamity is beyond any description, I, therefore, shall let the curtain
drop, and leave the consideration of this thrilling subject
to some one else. I am at least satisfied that we are now
having the true fulfilment of the Divine prediction of
our Saviour in Matthew xxiv, 7 ; St. Mark xiii, 8 ; St.
Luke xxi, 25. And their shall be signs in the sun, and
in the moon, and in the stars ; and upon the earth distress
of nations, with perplexity ; the sea and the waves roaring ; men's hearts failing them for fear, and for looking
after those things which are coming on the earth."

1. The clouds at length are breaking,
        The dawn will soon appear,
    And signs, there is no mistake in,
        Proclaim, Messiah's near.

2. Awake, awake, from sleeping,
        And let your works abound ;
    Be watching, praying, seeking,
        For soon the "Trump will sound."

3. Awake ye slumbering virgins,
        Send forth the solemn cry ;
    Let all the saints repeat it,
        The Bridegroom draweth nigh.

4. Let all your lamps be burning,
        Your loins well guarded be ;
    Each longing heart preparing,
        With joy thy face to see.

Brethren, while we sojourn here,
Fight we must, but should not fear ;
Foes we have, but we've a friend,
One who loves us to the end.

Forward then, with courage go,
Long we shall not dwell below ;
Soon the joyful news will come :
Child, your father calls—Come home.

The ways of religion true pleasure afford,
No pleasures can equal the joys of the Lord ;
Forsake then, the world and escape for thy life,
And look not behind you, remember Lot's wife.

Our next chapter, reader, will give a brief account of
some of the greatest fires of the last three years.

CHAPTER VIII.

FIRES.

Boston, Nov. 1, 1867—Two large shoe manufactories
in Southboro, belonging to Messrs. John Hart & Co., shoe
dealers, at 114 Pearl St., in this city, were destroyed by
fire on Tuesday morning. Loss, $100,000.

Steamboat Explosion.—Cincinnati, Jan. 30, 1867.
The steamer Miami, which exploded on the Arkansas
River, had about 250 passengers on board, among whom

were 91 men of Company 13, of the 13th United States Cavalry. The explosion was of such force as to rend the cabin floor asunder and let every person down who was in the front part of the cabin. 150 persons perished.

ANOTHER EXPLOSION.—Cincinnati, Jan. 31, 1867, The steamer Missouri, which exploded in the Ohio River had 120 persons on board. The latest information from Eranswelle places the loss of life at 100. The Missouri was valued at $100,000 and went all down.

FIRE AT BUFFALO, Jan. 20, 1868—About half-past 10 o'clock, last night, a fire broke out in the rear of Altman & Co.'s five story brick clothing store, destroying property to the amount of $300,000.

From the New York *Herald*, 18th Dec., 1867.—"In six months twenty steamboats, ten barges and three wharf boats have been destroyed by fire on the rivers of the west and south west. Total loss $1,197,000."

New York *Tribune*, March, 1867.—"The losses by fire in the United States, for several weeks, have averaged half a million dollars daily. The steamer Xerxes, from Cincinnati to New Orleans, with a full cargo, struck the sunken wreck of the Black Hawk below Mound City, took fire and was entirely consumed. The boat was valued at $60,000."

McGregor, Iowa, Dec. 8th, 1867.—A fire this morning consumed the Post Office, McGregor House, Mozart Hall, and some wooden buildings on the opposite side of the street. Loss estimated at $70,000.

Cincinnati, April 9th, 1868.—A fire at Edinjury, Ind., on Friday, destroyed the Johnson House, Dembert & Co.'s woolen factory, and Bickett & Co.'s distillery. Loss $60,000.

Ashland, Pa., April 9th, 1856.—There was a tremendous fire in this town last night. One whole square was destroyed. Loss $70,000.

From the Boston *Traveller*, March, 1867.—"The extensive foundry and locomotive works, formerly known as the Boston Locomotive Works, was nearly destroyed by fire on Saturday night. Loss, $75,000.

On Saturday night a fire broke out at the furniture store of W. W. Serving, Chicago. Loss, $100,000.—*Herald*.

New York, Jan. 21st, 1867.—" The storage ware-house of Miller & Conger, was destroyed by fire last night. Loss, $500,000. A destructive fire occured in Louisville, Kentucky, on Friday night, destroying the establishment of Newcombe, Buchannon & Co. Loss, $200,000."

Waterloo, Ind., Feb., 1867.—The Union Flour Mills were destroyed by fire this morning. Loss, $20,000.

Chicago, Feb., 1867.—The St. Charles Paper Mill, at St. Charles, Ill., was destroyed by fire on Sunday night. Loss $110,000.—*Tribune.*

New York, Feb., 1866.—Last night the Bleeker Street and Fulton Ferry Railroad Companies stables were destroyed by fire. Loss, $40,000.—*Herald.*

Albany, Feb., 1866.—A fire this morning destroyed on First Street, property to the amount of $45,000.

BURNING OF A SHIP AT SEA, Feb. 1866.—London *Times*: The Maggie Atkinson, of Shields, from Tuticorin, with a cargo of cotton valued at $250,000, was totally burned on the 13th, in lat. 25 N., lon. 39 W. The following table gives the losses by fire in the United States, from 1855 to 1864, inclusive, at $171,449,000.

GREAT FIRE AT VALPARISO, Nov. 1866.—Destruction of property by fire to the amount of $3,000,000. The village of Lima was visited by a destructive fire on Friday morning; three blocks burned. Loss $20,000. The City Hotel, Madison, Wis., was burned, Dec. 14th. Loss, $25,-000.—*Tribune.*

RACINE, Wis., Jan., 1866.—A fire this morning destroyed the Racine House block, the Titus block, the Telegraph Office, and other buildings.—Loss $100,000.

New York, March, 1867.—A fire broke out this morning in the cotton storage of E. C. Johnson, No. 4 Bridge St., and destroyed property to the value of $300,000.

St. Louis, Oct., 1867.—Lewis Leith's vinegar factory, on the corner of Washington street and Franklin Avenue, was burned this morning. Loss $40,000.

Battle Creek *Herald*, March, 1866.—The losses by fire throughout the United States for the six months just ended, exceed those of the whole of any previous year. Reckoned at $45,000,000.

Davenport, Iowa, Dec., 1867.—A fire last night destroyed property to the amount of $50,000.

N. Y., *Tribune*, 1866.—An extensive fire occured on Saturday night on the corner of St. Charles and Canal streets, and destroyed property to the amount of $60,000.

New Haven, Conn., Jan., 1867.—A fire broke out here this morning in Wenhall's extensive carriage and car factory. Loss $50,000.

N. Y. *Tribune*, Feb., 1867.—A fire this morning destroyed the City Assembly Rooms, the American Theatre, and several stores in Broadway. Loss $200,000.

St Louis, Mo., 1867.—The tobacco factory of Migers & Drummond, Alton, Ill., and one or two adjoining stores, were burned last Saturday. Loss $30,000.

From the Battle Creek *Herald*, 1866.—The past two years have been marked by a greater number of serious conflagrations in every part of the country than had ever been known before. Scarce a town of any considerable size has escaped, but none has suffered so tremendously as the beautiful city of Portland, the second maritime town of New England. More than one-half of its buildings were destroyed. Among them were five churches, its beautiful City Hall, all its bank and newspaper offices, its manufactories and stores, and its great hotels: The loss is said to exceed in value $10,000,000, and there are losses in such cases which cannot be estimated at a money computation.

New York *Tribune*, 1867.—The recent fire at St. Louis. The loss by the steamboat fire at St. Louis, Feb. 26th, including the cargoes of the Dictator and Luna, and freight burned on the levee, is estimated at $750,000. The Mansion school house was burned at 2 o'clock this morning. Loss $40,000.

TERRIFIC EXPLOSION.—A terrific explosion occured on Tuesday evening at the furnace of J. & H. J. Weilly, in Middleton, resulting in the complete destruction of the furnace, the death of five men, and the wounding of six others. Eight boilers were in the furnace, one was raised through the building, carried 500 yards and lodged in the Pennsylvania canal. The other boilers were scattered in every direction, some of them passing through houses and other buildings. A portion of one boiler was hurled through a room in which two women were lying sick, but missed them. The bridge over the Union canal was carried away. Many of the buildings in the vicinity were more or less shattered by the fragments, and the whole town was shaken. The loss exceeds $500,000.

St. Louis, Feb., 1866.—The steamers Luna, Leviathan, and Peytona, were burned at the levee, together with a large amount of freight. The loss cannot be ascertained as yet, but it will at least reach $500,000.

The New York *Times* says that it has been estimated upon good authority, that the losses by fire, during 1866, in the States, including the Portland fire, exceeds $15,000,-000.

Philadelphia, Jan., 1866. —A very destructive fire broke out here this morning, the 2nd, in Chestnut street. The buildings, 607 Chestnut street, occupied by Harnden's Express Co., Nos. 603 and 605, by Rocklill & Wilson, wholesale clothiers, and 609 by Perry & Co., clothiers, were almost entirely destroyed. The walls fell into the street and some of the firemen were injured. Loss, $100,-000.

From the Weekly *Globe*.—Great fire in Quebec. It lasted 13 hours. 2,500 houses destroyed. 18,000 persons rendered houseless. Several lives lost. The estimated loss nearly $3,000,000.

St. Louis, April, 1867.—The steamers Major Effie Deans, Nevada, Fannie Ogden, and Frank Bates were burned at the levee this morning. The boats belonged to the North American Fur Company, and were heavily laden with supplies for their trading posts. The loss on the boats and cargoes, $525,000.

New York *Tribune*, July 19, 1866.—The rope factory of Henry Lawrence & Sons, Williamsburgh, was struck by lightning yesterday, four boilers exploded, two of which were driven through a brick wall to a distance of 600 feet, destroying several large trees in their course. The other two passed through a brick stable, which was completely destroyed; one man was killed. The damage is over $20,-000. The Presbyterian church, on Fran n Avenue, Brooklyn, was damaged by lightning. Two persons were killed in the streets of Brooklyn. The cotton seed oil factory of the N. Y. Oil Co., 19th street, and the Primary

School House adjoining, were destroyed by fire last night.
The macaroni factory of Mr. Billings was damaged. Loss
of those, $50,000.

GREAT CONFLAGRATION AT DETROIT.—The most destructive conflagration that has ever afflicted our city, occurred last evening at an early hour, causing a loss, the
enormous amount of which will reach, without exagger-
ation, the sum of $1,000,000. The freight depot and shed
of the Central Railroad, stored with valuable commodities
representing capital of all portions of the country, were
totally destroyed, crippling the powers of one of the most
important and enterprising corporations of the West, and
inflicting blows upon our business interests, which, if not
amounting to total paralysis, will seriously retard their
development. The disaster is a momentous one, and we
chronicle its details with a regret that we have never before been compelled to feel at any merely local calamity.

Oswego, Dec., 1866.—The steam saw mill owned by
Chandler & Co., Wilmot, situated in the cove on the East
side of the river in this city, was consumed by fire last
night. The schooner J. L. Crocker had her masts and
bowsprit destroyed. The total loss is over $30,000.

Philadelphia, Ont., 1866.—The large new five story
grist mill of Jacob Kelne, near Phickemin, Somerset county, N. J., was destroyed by fire this morning; a quantity
of grain was destroyed also. The loss is over $25,000.

N. Y., Feb., 1867.—Yesterday afternoon, Hillyer &
Sons' drug mills, Jersy City, were destroyed by fire. Loss
$20,000.

New Haven, Ct., Feb. 9th, 1866.—Last night a large
tobacco warehouse in New London, Conn., containing
some $20,000 worth of tobacco, was destroyed by fire, and
all lost.

St. Louis, Feb. 12, 1867.—The car-house and repair
shop of the Ohio and Mississippi Railroad, opposite this
city, was burned this morning together with sixteen cars.
Loss about $40,000.

From the Globe, 1866.—We have received intelligence
from a correspondent at Elora, that at the early hour of
three o'clock on Saturday morning a fire baoke out at the
Elora Mills and Distillery, owned by J. M. Frazer, Esq.,

and in less than an hour the whole of the premises were completely destroyed. The loss is $20,000.

Tribune, 1867.—The United States bonded warehouse on Dickerson street wharf, containing 4,000 barrels of coal oil, belonging to various parties. Loss $96,000

Pithole, Oct. 9th, 1866.—The most destructive conflagration that ever took place in the oil regions occurred here at one o'clock this morning, which terminated without loss of life. The tanks of the largest flowing well in Pithole, known as Grant's, containing 4,000 bbls. of oil caught fire, spreading destruction in all directions. Thirteen derricks and engine houses on the following leases were destroyed: No's. 16, 17, 18, 19, 20, 21, 22 and 23, and on Hormder farm, and No's. 24 and 25 on the Morris farm, covering a surface of five acres. The total loss will reach $150,000.

DESTRUCTIVE FIRE AT OTTAWA.—From the Ottawa *Citizen*, Oct. 15, 1868.—We had a large fire in this city last Sunday morning, destroying property belonging to various parties, to the amount of from $20,000 to $50,000.

Tribune, 1866.—N. Y. Aug. 20.—A fire at Jersey City yesterday, was one of the most disastrous that ever occurred in this vicinity. It burned from eight a. m. till nightfall. It broke out on the schooner Alfred Barrett, lying at the oil pier on Jersey shore. She was laden with oil, and was to have started for Boston this morning. On her catching fire she blew up, and the flames spread with great rapidity to the adjoining piers and vessels. Two piers were totally destroyed and one partially. Thirteen lives lost and several persons injured. The loss of the property was estimated at about $2,000,000.

TREMENDOUS FIRE IN CARIBOO.—The principal town in ashes. One hundred houses burnt. Loss over one million dollars, ($1,000,000.)

From the British *Columbian*, New Westminster, Sept. 23, 1868.—It becomes our painful duty to chronicle a fire in comparison with which all that have previously happened in this Colony are as a drop to the bucket. On Wednesday, the 16th inst., about 2 p. m , a fire broke out in Barry's saloon, Barkerville, and before 5 o'clock the entire town was one mass of smouldering ashes. When it is understood that Barkerville was the principal town in.

Cariboo, the depot in fact, for that entire region, some idea may be formed as to the extent of the disaster. The loss is variously estimated from $1,000,000 to $2,000,000. And if the cost, or, indeed, the market value of the building be taken into the account, we fear that the higher figure would not be found out of the way. With winter so near, and no time to replace much that has been burned, it is to be feared that the mining interests must suffer, and that many who would otherwise have remained in Cariboo, will now have to leave the mines. Unquestionably there must be a very large quantity of goods in transit between the seaboard and Cariboo, for shipments were never so heavy as they have been during the past six or eight weeks. But, view the matter in the most favorable light, one cannot avoid the conclusion, that to have some two million dollars worth of property swept away in a moment, and so large a population left not only penniless but houseless at this late season, is a blow sufficiently heavy to make itself felt throughout every district and in every interest.

My kind reader I shall make one quotation more before I bring this calamatous subject to a close.

From the Philadelphia *Ledger*, of Nov., 1866, who says : —From May 15th to October 15th, (five months) there were 68 railway accidents, and in the whole year to date not less than 100. 'Thirty millions would be a modest reckoning,' says the writer quoted above: 'For the loss involved in demolished lives and property. Probably 300 have been killed, and more than 600 injured in every shape, during the year. 65 disasters by steamboat explosions and shipwreck are also noted. Several of these catastrophes which crimsoned river and ocean, far and wide, and remembered in all their awful poetry, and helpless agony and terror. From the Sultana, some 1,-200 found a muddy grave in the Mississippi, 400 went down in the burning ship Nelson, 250 with the Brother Jonathan, and 100 in the Pewabic, 3,000 lives, it may be reckoned, have been lost between April and September.' In addition, may be added, a great variety of miscellaneous casualties from incavations, burnings, crushings, shootings, cuttings, and particularly explosions, from which latter cause an enumeration of only five cases gives a list of killed and wounded, reaching nearly 2,000. The same writer says :—We still hear the most heart-rending accounts of disasters at sea, caused by the recent storm in the Gulf of Mexico and on the coasts of Carolina and Florida. At least 100 vessels have been wrecked and many lives lost. Conflagrations must be added to the catalogue of items in the area of ruin. 155 fires between April and October 15th are minuted. A brief table of losses is subjoined.—

The loss by unenumerated fires.................... $30,000,000
Burning of the Government Works in Tenn... 10,000,000
Warehouse conflagration in New York......... 3,000,000
Other fires, also 1,500,000
In the same period, fires in Canada...... 1,500,000

Grand total in six months....................$45,000,000

My intelligent reader will, I think, admit that the fires of late is unusual in the extreme. The Prophet Amos, iv, 9, 13, says :—" I have smitten you with blasting and mildew ; when your gardens and your vineyards and your fig trees and your olive trees increased, the palmerworm devoured them ; yet have ye not returned unto me saith the Lord. I have sent among you the pestilence after the manner of Egypt." See Exodus, 7th, 8th, 9th, 10th and 11th chapters, and I think the next three years will fulfil this verse in Amos." Verse ii, " I have overthrown some of them as God overthrew Sodom and Gomorrah." See Genesis, xix, 23, 24, and 25. Reader, the reason I refer you to this last passage is, that I expect according to Amos' prediction. some of our ungodly cities before three years will be burned, as Sodom and Gomorra were in the time of Lot.

O give me the Bible—the statutes of heaven,
  Its great constitution I know to be pure ;
All ten of its precepts in justice are given,
  And all is divine and unalterably sure.

I know when I read them in love they were blended,
  Nor one disannulled since the time they were framed,
No foul legislation has ever amended,
  One jot or one tittle that therein is named.
The old-fashioned Bible, the dear blessed Bible,
  The family Bible that lies on the stand.

Tho' thousands have written a substitute for them,
  To sway over others the sceptre and sword,
Yet even unaltered these laws lie before them,
  Unchanged and immutable—word of the Lord.

Then give me my Bible and let me obey it,
 Instead of the statutes and doctrines of men
Aside for a moment, forbid I should lay it,
 To listen and argue for dogmas again.
The old fashioned Bible, the dear blessed Bible,
 The family Bible, that lies on the stand.

The next and last chapter of this book I shall introduce
by way of a dialogue between the reader and writer.

CHAPTER IX.

The " Signs of the Times."

I hold that the usual fires, floods, hurricanes, pestilence,
famine, earthquakes and crimes, a sure sign of the coming of the day of judgment. See Matthew xxiv ; Mark
xiii and St. Luke xxi, chapters. Read also Amos iv, 9, 10,
11 ; Haggai ii, 16, 17. In those chapters are mention
made of what was to be the tokens of the coming of
Christ. The Reader—But how can it be known definite
about the coming of Christ? Writer—First, I argue that
God has not altered, but is the same yesterday, to-day and
for ever. God informed Noah of the flood, and the Apostle
Peter, in his second epistle, chapter ii and verse 5, calls
Noah a preacher of righteousness. Now, I would ask the
reader candidly what did Noah preach? Noah was informed by Jehovah that he was to bring a flood
upon that ungodly generation in which Noah was among.
And to my mind, reader, Noah, as every godly man would
do, went to preach the truth of the flood to those that were
around him; otherwise, God could not be justified in
bringing the flood without due warning. Reader—This
truly is admitted, but what gain is the admission to you?
Writer—Every gain. For surely the antedeluvian age
was not more interested in their salvation than we are,
and if it is as we say, God is not partial, and surely he is
not; then, they got notified of the flood, why not we of
the end of time ? And very beautifully this is illustrated
by our Saviour in Matthew xxiv, 37, 38. Where Christ
says, that as it was in the days of Noah, so should it be in
his coming. Reader—Yes, but Christ said " of that day and
hour knoweth no man, no, not the angels of heaven, but
the Father only." Writer—Christ said, I believe the very
words about 1800 years ago ; but, Christ did not say that

no man, "or angels, nor He himself, ever should know the day and hour, for, if God is not changed, them of his people that is living immediately before the coming of Christ will know, as Noah and Lot did. See Genesis vii, 4 " For yet seven days," (here is definite time given, " seven days,") never forget this reader, in your arguments about the day and hour. Read also, Genesis xviii chapter, on the destruction of the cities of the plain. Remember also, what kind of people was to be on the earth at the last days. Peter says, " there shall come in the last days scoffers, and saying where is the promise of his coming ?" Now, in all earnestness, 1 would ask the reader how could the scoffer speak thus if there was no one setting forth the promise of His, Christ's coming; it would be a perversion of terms, which I shall not admit to be existing in God's word. And again, hear what our Saviour says in Matthew xxiv, 45. " Who then is a faithful and wise servant, whom his lord hath made ruler over his household, to give them meat," (the truth of the signs that 1 gave in the preceeding verses) " in due season ?" " But and if that evil servant shall say in his heart, my Lord delayeth his coming ; and shall begin to smite his fellow servants," here is one class preaching the truth of the coming of Christ, the other class is smiting those, not, I presume, by the fist, but by a slandering tongue ; but yet the time of Christ's coming was to be preached prior to his coming. or the following scripture would be without meaning. See Matthew xxv, 5, " While the bridegroom" (Christ) " tarried, they all slumbered and slept " Reader—Who slumbered and slept? Writer—Simply those that were aroused by the preaching of time in 1843 and 1844. Nothing can be said to tarry, except a set time is given first, then, that is passed and gone, comes the sleeping and slumbering spoken of in the quotation above. Listen what Paul says in Hebrews x, 35, 36, 37, 38, and 39 verses. " Cast not away therefore your confidence, which hath great recompence of reward. For ye have need of patience, that, after ye have done the will of God," (that was the time) " ye might receive the promise ; for yet a little while, and he that shall come " (Christ) " will come, and will not tarry " The Reader—We should not meddle with time, those periods that some of late years have been preaching about, are the secrets of God. Writer—Moses says. Deuteronomy xxix, 29, " The secret things belong

82

unto the Lord our God : but those things which are revealed belong unto us and to our children for ever."
Therefore, my kind reader, whatever we find in the scripture of truth, is our inheritance by promise, and Paul says,
in second Timothy, 3rd and 16th, that, " All scripture is
profitable." I say nothing can be of any profit unto us,
except we comprehend it. Peter, also, gives his testimony
in regard unto the writings of the prophets, and says,
" That we do well to take heed, as unto a light that shineth
in a dark place," and this is my motive reader, in bringing
those blessed testimonies to your notice. The Reader—All
very well, but we should leave the prophetic periods altogether alone. Let us, in answer to this, your opinion,
see what says the scripture, see Matthew xxiv, 15. Here
comes the language of our blessed Redeemer, hearken,
" When ye therefore shall see the abomination of desolation spoken of by Daniel the prophet, stand in the holy
place, whoso readeth let him understand." This language, my kind reader is rather opposed to your idea of
those things, for Christ wants us both to read, and understand what we do read. And suppose we would admit
for a moment, that time, as given in the word of God, is a
secret ; hear the Psalmest in the xxv Psalm and 14 verse,
who says, " The secret of the Lord is with them that fear
him ; and he will show them" (that fear him) " his covenant" (the truth.) Prover. ii, 32, " For the froward is
abomination to the Lord : but his secret is with the righteous." Chapter iv, 18, " But the path of the just is as a
shining light, that shineth more and more unto the perfect
day." Chapter vi, 23, " For the commandment is a lamp;
and the law is light ; and reproofs of instruction are the
way of life." Chapter xxvin, 4, 5, " Evil men understand
not judgment : but they that seek the Lord understand all
things," time, like every thing else, as that is on record.
Again, see Eccles. viii, 5, " Whoso keepeth the commandment shall feel no evil thing : and a wise man's heart dis-
cerneth both time and judgment." See Paul, in Acts xvii,
31, " Because he" (that is God) "appointed a day, in the
which he will judge the world in righteousness." Daniel
xii, 10, " Many shall be purified, and made white, and
tried ; but the wicked shall do wickedly : and none of
the wicked shall understand ; but the wise shall understand." Settle upon this, reader, as the yea and amen.
Hear again, and understand, what the Prophet Habakkuk

saith chapter second, first and second verses. "I will stand upon my watch, and set me upon the tower, and I will watch to see what he will say unto me; and what I shall answer when I am argued with. And the lord answered me, and said write the vision. and make it plain upon tables, that he may run that readeth it, for the vision is yet for an appointed time, but the end it shall speak, and not lie, though it tarry," (here is the time appointed in 1843 and 1844, and the tarrying ever since) " wait for it; because it will surely come, it will not tarry." The Reader—The time was preached and failed before, and is it not likely to do so over so many times again. My answer to the foregoing is, that the very generation that have had the time preached unto them, shall not pass away, until Christ will come. Let us see again what the prophet Ezekiel says on this point, Chapter xii, 21, 28, " The word of the Lord came unto me, saying, son of man, what is that proverb, or by-word, that ye have in the land of Israel. saying; The days are prolonged, and every vision faileth." Here my kind reader, is a perfect notice taken of the arguments of the scoffers, they say that every vision faileth, which proves to a demonstration that time was preached, and failed at times. which I cannot deny, nor do desire to do so ; but here again, in the quotation above, " Tell them therefore, thus saith the Lord God : I will make this proverb to cease, and they shall no more use it as a proverb in Israel ;" (or among the sects) but say unto them, the days are at hand, and the effect of every vision." This then my kind reader, is the very thing I want to inform my fellow travellers to the judgment seat of Christ, that the " days," are at hand, and the " effect of every vision." The Reader—Well, but you were to show us, that it should come in the generation that heard the time preached. I shall endeavor so to do, hear the same prophet, in verse 28, " Therefore say unto them," (hear it reader) "Thus saith the Lord God; There shall none of my words be prolonged any more," (or delayed if you will,) " but the word which I have spoken shall be done, saith the Lord God." Do you believe this reader, see again verse 25th, " For I am the Lord : I will speak, and the word that I shall speak shall come to pass; it shall be no more prolonged ;" (or delayed) " for in your days," (here is the sealing) " in your days, O rebellious house, will I say the word, and will perform it saith the Lord God. The veto, is on it my kind reader, that in the days of those that have heard the time preached, Christ will make his appearing, and but few looking for him, or loving his appearing.

1. O, come, come away, for time's career is closing ;
   Let worldly care henceforth forbear ; O, come, come away,
   Come, come, our holy joys renew, where love and heavenly
       friendship grew ;
   The spirit welcome you ; O, come, come, away.

2. Awake ye, awake, no time for reposing ;
   The Lord is near, breaks on the ear ; O, come, come away,
   Come, come, where Jesus' love will be,
   Who says, I meet with two or three ;
   Sweet promise made to thee ; O, come, come away.

3. With joy I accept the gracious invitation ;
   My heart exults with rapturous hope ; O, come, come away ;
   When Jesus comes, O may we meet
   A happy throng at his dear feet ;
   Our joy will be complete ; O come, come away.

4. Come where sacred songs the pilgrim's heart is cheering,
   Come there, and learn the power of prayer ; O come, come away ;
   In sweetest notes of sympathy
   We praise and pray in harmony ;
   Love makes our unity ; O come, come away.

5. Night soon will be over, and endless day appearing ;
   Away from home no more we roam ; O come, come away ;
   And when the triumph of God shall sound,
   The saints no more by fears are bound ;
   We own our Jesus crowned ; O come, come away.

6. O come, come away, my Saviour, in thy glory ;
   Thy kingdom come, thy will be done ; O come, come away,
   O come, my Lord, thy right maintain,
   And take thy throne and on it reign,
   Then earth shall bloom again ; O come, come away.

My very indulgent reader, that I may not be too tedious unto you, I shall return to our former conversation, about time. The reader—There are several periods in God's word, which if we knew their commencement, there could be no mistake in their ending ; but that is the question. I want my reader, to bear in mind one thing ; there were several eminent scholars endeavored to give us the dates of events which we call chronology. Bishop Usher's chronology is the one we have in our Bibles, that which we call King James' translation ; and his chronology is 25 years ahead of Clinton, Rev. R. Shimeall, Rev. E. Elliott and Haines; it is proved itself to be incorrect, for if it had been correct, we would now be in the great scenes, beyond the bounds of probation. Hence, with your kind permission reader, I will give the two chronologies, Bishop Usher's, and those other eminent scholars. First of all, then, is Moses seven times in Leviticus, xxvi, 24, 28 verses; where the reader can see the time that Jehovah was to punish his people, if they would not keep his commandments, which they did not. Those seven times are equal to seven years, prophetic, or, seven times twelve is eighty-four, multiplied by thirty, the number of days in each month, will give us the whole length of time that was to be occupied in the persecution of God's people, from a certain given point of time, until they should be delivered at the end of this dispensation. In all two thousand five hundred and twenty days, or, so many years. This punishment commenced, as you can see by referring to second, Chronicles xxxiii, 1, 2. According to Bishop Usher's chronology in the year B. C. 677, which if you will deduct from the great period of 2520, will bring us to Mr. Miller's time, 1843. But according to those eminent scholars mentioned above, would bring us to the year 1868, A. D. But the reader will say 1868 is also past; hence, the whole chronologers have failed, and what then? We must not in our anxiety, to confute the time and its advocates, forget, that one of the Popes altered the Christian Era, 4 years ahead.

The sum stands thus................................. 2520
Moses seven times, or 2520 years, by Usher's
    commenced B. C.,.................. 677

And as a simple rule ended in A. D.,......... 1843
                              25

To which add the 25 years difference in
    chronologers .... 1868 A. D.
And again to this we must add the four years 4
    that the Christian Era has been
    set ahead brings us unto the year
    of our Lord.................. .... 1872 A. D.

full time, or the beginning of 1873. Spring Equox, or Exodus. This is the way I view those periods, I do not know that the event will then come; but I believe it from my heart. "Faith is the assurance of things hoped for," the evidence of things not seen, for example, the coming of Christ in 1873, but I believe it, and I will proclaim it unto all the world. The Reader—is this all the evidence of the coming of Christ, or, is there more proof? and if there is, does it clash with that given above? I shall briefly touch upon the 2300 days, or years, given in Daniel chapter viii. 13, 14 verses, which reads as follows: "Then I heard one saint speaking, and another saint said unto that certain saint which spake, How long shall be the vision concerning the daily sacrifice, and the transgression of desolation, to give both the sanctuary and the host to be trodden underfoot? And he said unto me, Unto two thousand and three hundred days; then shall the sanctuary be cleansed." The Reader—are those days to be understood as literal time, the 2,300 days, so as to make only six years, and a little over four months, at which end the sanctuary was to be cleansed? In answer to the above, I would say, that God has left on record a rule, by which we can know the end thereof. First, we must apply the literal rule, that one day means the revolution of the sun in twenty-four hours. Second, the figurative rule, that one day in certain instances, means a year; where that construction of the word is justified. I suppose no one will ask me to prove my first rule; but the second I shall have to substantiate, by thus saith the Lord. Please to refer to Numbers xiv, 34. Ezekiel iv, 5 and 6 verses. In those passages God says by the prophet that he has given "each day for a year." Therefore, this is the rule we shall apply to the 2,800 days that they mean just so many years. Because we find the date this vision was given to the beloved Daniel, to be according to Usher's chronology 553 years before Christ, and of course, would only extend a little more than six years beyond that time, and as the sanctuary was not then cleansed, according to promise, nor is yet cleansed, we look for the latter rule. When Daniel had the above vision, his mind was troubled to know its meaning, and God sent his Angel to inform him how, and where to commence the 2.300 years. Daniel ix, 25, "know therefore and understand, that from the going forth of the commandment to restore and to build Jerusalem unto the Messiah the Prince." (or Christ,) &c. Here is the starting point from which to commence the 2,300 years, but as there was three such commandments given, we are forced to take the last one, (i e) that given in Nehemiah by Artaxerxes, Nehemiah 1, 4, in the year B. C. 432. By taking either of the two first commandments for restoring Jerusalem, namely, the one by Cyrus, in the first year of his reign, (Ezra i, 1,) B. C. 536, or that by Darius in his second year, (Ezra iv, 24,) B. C. 520, as the commencement of the seventy weeks, the ending in one case would be forty-six, and in the other 30 years before Messiah was born. The seventieth week was devoted in confirming the covenant with none but Jews, and in the middle of it Messiah was cut off. He confirmed the covenant with Jews only during the three and a half years of his personal ministry, and for the remaining half of the week, or three and a half years, his disciples did the same, until Peter opened the church to the Gentiles by the baptism of Cornelius and his house. The Reader—we cannot be sure that the great period of the 2,300 years begins with the lesser period of seventy weeks, or the going forth of the last commandment for restoring Jerusalem, because it is not expressly said it should then begin. I answer, we have seen that it could not possibly have begun either when Daniel saw the vision, or at giving of either of the two first commandments, for then all the events mentioned to transpire within the 2,300 years, and this too is applying our last rule must have been completed more than fifty years since. For the date of Daniel's vision, B. C. 553, being deducted from the 2,300 years, leaves A. D. 1747 as the end; the date of the first commandment, 536, being deducted, leaves A. D. 1764, and that of the second command, 520, being deducted, leaves, A. D. 1780. The third and last decree for finishing the city, given B. C. 432 therefore furnishing the only light that I can see in God's blessed word, for the commencement of the 2,300 years, if there is any; I solicit the world to show it, until this is done I am forced to adopt my own rule;

This is it......	2,300 years
Commenced in the year before Christ	432
In subtracting the 432, from the 2,300 years	——
it leaves...............	1868 A. D.
To which we must not neglect to add the 4 years of Rome.................	4
	——
	1872 A D.

which will bring us to the year 1872, full years, or to the Exodus of 1873 in the spring; this is my foot-hold, here I stand. Furthermore, it is evidently as necessary for the church to know the commencement of the 2,300 years as it was to know that of the seventy weeks, which were a part of them, or that of the 430 years of oppression in Egypt, otherwise there would be no limit of time presented to us within which to bring the events of time, and the great end of prophecy would be defeated, which is to warn the church of coming events, and the neighborhood of their appeals, and to have a true ground of judgment on which to convict her of apostasy and unbelief in turning aside from the prophetic word, after her own wisdom. The Reader—but I thought prophecy was left a mystery, that we cannot understand. Well, my kind reader, prophecy is given to keep alive expectation, that when the church sees the premonitary "signs coming to pass, she may lift up her head and know that her redemption is drawing nigh." St. Luke xxi, 28. Dates are given to the prophetic periods, and signs whereby some of them may be ascertained during the progress of fulfilment, to inform the wise when they begin and when they end. If such is not the design of dates and periods as well as signs, it is hard to conceive what is their object in being given. If the prophetic announcements were thrown out into the limitless void of time indefinite, as they are in all the prophets except Daniel and St. John, the things predicted could never appear as things to fall within the experience of any particular generation of men, as tangible realities, and we should float down the stream of time, without chart or compass or waymark; and hence, that prophecy would be to us a light shining in a dark place, God was graciously pleased to furnish us with the chronology of prophecy by Daniel and John. The giving of periods were little else than a mockery if the means of ascertaining their beginning and ending were not also given; hence, the giving of dates and way-marks is just as necessary as the giving of periods. God was careful to furnish these dates and way-marks so early in the course of events foretold, as that his people should have ample time of preparation for the approach of the more important ones. It was so of the seventy weeks, or 490 years, and the thing itself shows that it is as needful to know what period the Lord's second coming to judgment is apprehended, as within what period his first coming and suffering were com-prehended, inasmuch as a great and fearful judgment is the announced attendant upon each event. Each period is a definite period, having a beginning and an ending, and containing a given number of years. The Jews knew when the lesser period began and when it ended, for a considerable time before it did end; therefore, I contend there is no reason why we of this day should not in like manner have the means of knowing the time of the open-ing and close of the 2,300 years, which evidently brings us all to the end of the Gospel dispensation. But the Jews as a church and nation, rejected the evidence of prophecy which went before the Messiah, and there are causes urging us to reject the evidence of dates, and disbelieve and deny the prophecy according to the example of the Jews, these are therefore reasons to call upon us, to have faith in the prophecies, for without faith we cannot please God. There were mighty events connected with the first advent of Christ, of which God was pleased to warn the Jews before hand; there are events still more stupend-ous connected with his second advent, events of great in-terest, both to Jews and Gentiles, of which both are dis-tinctly warned in all the prophets; and I repeat it, there is the same reason why all parties concerned should have the means of knowing the very year of the completion of the 2,300 days, (years) in order to be prepared for these mighty events, that there was for the Jews to have the means of knowing the year of the ending of the 490 days, (years.) God judged the Jews, destroyed their city, and sent them into a long and painful captivity until the "times of the Gentiles shall be fulfilled," because, refusing to un-derstand the prophecies, they knew not the time of their visitation by their Saviour, hence, rejected him. And so also the mighty destruction about to fall upon this world, will come because of the same evil heart of unbelief, in re-fusing to believe the prophetic word declaring these things. Thus God hath measured off 2,300 years that he might know the truth. He gave us the death of Christ to seal and make sure the vision, just 486½ years from the commencement of the 2,300 years. The sum stands thus: as 486½ years reached exactly to the cross of Christ, so 18-13½ years more from the cross, will reach to the end of the vision, 2,300 years. We are passed the cross, and are closing up the last years of the 1813½, ought we not to

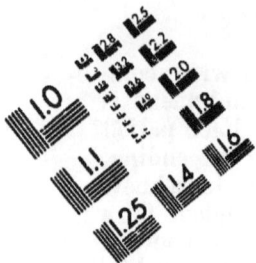

**IMAGE EVALUATION
TEST TARGET (MT-3)**

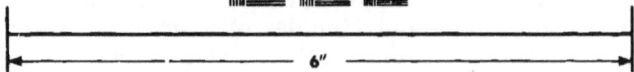

6"

Photographic
Sciences
Corporation

23 WEST MAIN STREET
WEBSTER, N.Y. 14580
(716) 872-4503

walk careful? We stand upon the verge of time, and the ending of a period of solemn importance. Every year, yea, every week or day we are to look for the crash of nations; war, famine, pestilence, tremendous fires as also very frequent awful floods, earthquakes, heavy and often; crime will be greatly on the increase, dry summers, failure hereafter of the crops. Soon and Daniel will stand in his " lot" or have his " inheritance," with the rest of God's people. Soon and the wicked shall be cut off from the earth. Soon the day of judgment will begin, solemn indeed. Are we ready for the solemn event? Have we repented of and forsaken our sins? Have we fled for refuge to lay hold on the hope set before us in the gospel? Have we made our judge our friend? Not a moment is to be lost. Soon the stone in the second of Daniel will smite the Roman image. Soon, as in Revelation, it will be said that " the kingdoms of this world will become the kingdom of our Lord and Saviour Jesus Christ. Soon and the door of mercy will be shut against an ungodly hypocritical church. Soon and the foolish virgins will say, " Lord, Lord, open unto us," but the answer is, " I know ye not, depart ye cursed." The Reader—is there nothing in the numbers that brings the periods more definite, and more intelligent to our understanding? We shall proceed to answer the above, by introducing the last periods that I shall speak from in this book; they are recorded in the xii chapter of the book of Daniel, beginning with the 7th verse, " And I heard the man clothed in linen, which was upon the waters of the river," (or stream of time,) " when he held up his right hand and his left hand unto heaven, and sware by him that liveth forever that it shall be for a time, times, and an half, and when he shall have accomplished to scatter the power of the holy people, all these things will be finished." The periods here spoken of, I understand to be and mean three and a half prophetic years, or in all 1,260 years. Those 1260 years, I also understand was the period that God's people was to be in the hands of Rome—which hands were cruel in the extreme—in this connection it does not say so; but, we shall refer you reader, to where the matter is made plainer. See Daniel vii, 23,28, speaking of the power of Rome, in the 25 verse, he says, " And he shall speak great words against the most high, and shall wear out the saints of the most High, and think to change times and laws; and they" (the saints) " shall be given into his" (Popery's) " hand until a time and times and the dividing of time." In all 1260 years. See again Revelation xii. 14. "And to the woman" (the christian church) "were given two wings of a great eagle, that she might fly into the wilderness, into her place, where she is nourished for a time, and times and half a time, from the face of the serpent" (Rome.) You see reader, that those periods was to be the persecuted time of Rome, against the people of God. Again in revelation xiii, 5 verse, the same periods are brought to view again; 'And there was given unto him" (Rome) "a mouth speaking great things and blasphemies; and power was given unto him (Rome to continue forty and two months." Revelations xi, 3 verse, " And I will" says God " give power unto my two witnesses," (the old and new Testament,) "and they shall prophecy a thousand two hundred and three score days, clothed in sackcloth." Here then reader it makes no matter how you take it, it ends the same way. That is, time, times, and half a time, which is three years and a half multiplied by 12, the number of months in the year gives us 42 months, that multiplied by 30, the number of days in each month, will give 1260 years, as sure as that one and one makes two. The Reader—how can it be comprehended, admitting all the above to be correct, when those periods commenced? I shall briefly answer the question above, and say, the 1260 years commenced in A. D. 538, when Justinian the Greek Emperor of Constantinople, constituted the Pope of Rome supreme head, over all the churches in the East, by giving him three kingdoms the Ostrogoths, Vandels and Heruli, the civil and ecclesiastical power, was at this time conferred on the Pope of Rome, by Justinian. I shall presently prove my position to be correct; because, beginning with the 1260 years in 538 A. D., you can see by adding the two numbers together, they will bring us down the stream of time through the dark ages, to the noted year 1798. There, you will find by the order of Bonaparte, Popery dethroned. And notwithstanding Popery is still Popery, yet, they have not the power they use to wield and enjoy, nor will not thank God, until within fifteen literal days of the end of this age. Bear in mind my intelligent reader, that in the same connection where we found our time, times, and half a time, exists two more periods, the last one bringing us to the Resurrection of Daniel, and if Daniel will rise at the end of the last period, so will all God's saints, at identically the same time. See Daniel xii, 10, 13, "Many shall be purified, and made white, and tried; but the wicked shall do wickedly: and none of the wicked shall understand; but the wise shall understand. And from the time the daily sacrifice shall bo taken away, and the abomination" (Popery) "that maketh desolate set up, there shall be a thousand two hundred and ninety days. Blessed is he that waiteth, and cometh to the thousand three hundred and five and thirty days." But Daniel is told to go away till the end would be, being assured that he would stand in his "lot at the end of the days." The Reader—where did those two last named periods begin, that is, the 1290, and the 1335 days, years? You recollect my kind reader, where we started with the 1260 years, those two last periods of course is in the same connection, and must have their beginning at the same place, namely,

In A. D................................... 538
The sum stands thus, 1260 begin in A. D....... 538
We will have them aside one another............1260

The time, times and half ended in A. D.......1798

We will follow our starting point................. 538 A. D.
With our next number 1290 years...............1290

This last period ended in A. D...................1828

We will still keep to our starting point......... 538 A. D.
With our third and last number..................1335 yrs.

Where will the resurrection take place but in 1873 A. D. This, therefore, is the end and substance of the whole matter that 1873 A. D., in the spring, ends every period, according to my knowledge and belief. On this rock I stand or fall, and I have no fear of the result. There is nothing but the most glorious harmony in all those blessed periods. Moses seven times, or, 2,520 years will bring us down to the spring of 1873. The 2,300 years bring us to the spring of 1873. And the 1335 years, from A. D. 538, will bring us to the Spring of 1873 Has the passing of set times demonstrated that the time will not be eventually known? Were I to waste time, ink and paper in answering this question, I should place myself in the same position as those who have recourse to this

argument. Now, all prophecy contains in its substance
both promise and threatening, addressed to the reason,
conscience, and understanding of all intelligent men, concerning things present and future, a scheme of things going forward in course of fulfilment, some part of which
concerneth every age of the world, yea, every year and
hour until the consummation of the whole. Both promise
and threatening are of the nature of prophecy, because
they concern the future touching both persons and things.
But in all earnestness, how shall the church act in regard
to God; promises and threatenings as yet unfilled without some definite and clear knowledge of things yet
future? I don't mean to say a perfect knowledge of all
things, times and circumstances, alone, will brighten the
path, for it might be pernicious to have a full knowledge,
but of some leading features, such as are to be seen in
the following example from 2nd Samuel, vii, where God
said to David, "Moreover I will appoint a place for my
people, Israel, and will plant them that they may dwell
in a place of their own, and move no more; neither shall
the children of wickedness afflict them any more, as before time, and as since the time that I commanded Judges
to be over my people, Israel, and have caused thee to rest
from all thine enemies. Also the Lord telleth thee that he
will make thee an house. And when thy days be fulfilled,
and thou shalt sleep with thy father, I will set up thy
seed after thee, which shall proceed out of thy bowels,
and I will establish his kingdom. He shall build an house
for my name, and I will establish his kingdom for ever."
Fully to interpret this passage, would be to interpret a
very large portion of Holy Writ. It is promise, and
threatening, and prophecy all woven inseparably into the
same web, and rightly to understand one requires the understanding of both the others. The very sight of the
words unavoidably carries forward the mind to the con-
templation of things yet future, and elicits an act of inter-
pretation, and points to certain definite things, and a certain definite spot of the earth for the theatre of the fulfilment of the promise to Israel, and to David, Israel's king.
But God says, "The stork in the heaven knoweth her appointed time, and the turtle, and the crane, and the swallow
observe the time of their coming, but my people" (christian professers) "know not the judgment of the Lord."
" For, as in the days that were before the flood, they were

eating and drinking, marrying and giving in marriage, until the day that Noah entered into the ark, and knew not until the flood came and took them" (the wicked) "all away—even thus shall it be in the day when the Son of Man (Christ) is revealed." And now my kind reader, I shall leave the subject with you in its present form, just as I see and believe, and have given you the evidence—as presented before us in the scripture of truth, that 1873, in the spring of that year, is the farthest point to which the 2,520, 2,300 and the 1335 years from 538 will allow us to extend them. My kind reader, I shall bid you farewell, hoping that you will give this thrilling chapter a serious perusal, and if you have a better light on this subject, than the one I have just given you, let us come and reason together, for it is the truth alone that shall make us free and wise unto salvation, and God shall have all the praise.—Amen,

Errata.

Page 5, 8 lines from top, for 33,332,333, read 31,554,059.
" 5, 9 lines from top, for 91,554, read 86,400.
" 5, 9 lines from top, for 3,730, read 3,600.
" 5, 10 lines from top, for second, read minute.
" 8, 6 lines from bottom, for 19, read 18.
" 11, 17 lines from top, for oppression, read oppressors.
" 15, 11 lines from top, for not read now.
" 16, the bottom line, for be read he's.
" 25, 12 lines from top, for women read woven.
" 27, 12 lines from top, for x, 30, read xxiv, 12.
" 36, 7 lines from the bottom, for crime, read come.
" 51, 12 lines from the bottom, for interests, intents.
" 88, 11 lines from the top, for appeals, read approach.
" 88, the bottom line, for apprehended, read comprehended.
" 89, 8 lines from the bottom, for he, read we.

1. How prone are professors to rest on their lees,
   To study their pleasure, their profit and ease ;
   Though God says arise, and escape for thy life,
   And look not behind you ; remember Lot's wife."

2. Awake from thy slumbers, the warning believe,
   'Tis Jesus that calls you the message receive ;
   While dangers are pending, escape for thy life,
   And look not behind you ; remember Lot's wife.

3. The first bold apostate will tempt you to stay,
   And tell you that lions are found in the way ;
   He means to deceive you, escape for thy life,
   And look not behind you ; remember Lot's wife.

4. How many poor souls has the tempter beguiled,
   With specious temptations how many defiled ;
   O, be not deluded, escape for thy life,
   And look not behind you ; remember Lot's wife.

5. The ways of religion true pleasure afford,
   No pleasures can equal the joys of the Lord ;
   Forsake then the world and escape for thy life,
   And look not behind you ; remember Lot's wife.

6. But if you're determine the call to refuse,
   And venture the way of destruction to choose ;
   For hell, you will part with the blessings of life,
   And then, if not now, you'll remember Lot's wife.

JAMES CALEB McINTOSH,

Bayfield, County of Huron, Ontario.

www.ingramcontent.com/pod-product-compliance
Lightning Source LLC
Chambersburg PA
CBHW021947190326
41519CB00009B/1171